Transparent Ceramics—a Theme Issue in Honor of Dr. Adrian Goldstein

Transparent Ceramics—a Theme Issue in Honor of Dr. Adrian Goldstein

Guest Editor

Yiquan Wu

MDPI

Basel • Beijing • Wuhan • Barcelona • Belgrade • Novi Sad • Cluj • Manchester

Guest Editor
Yiquan Wu
New York State College of
Ceramics
Alfred University
Alfred
USA

Editorial Office
MDPI AG
Grosspeteranlage 5
4052 Basel, Switzerland

This is a reprint of the Special Issue, published open access by the journal *Ceramics* (ISSN 2571-6131), freely accessible at: https://www.mdpi.com/journal/ceramics/special_issues/ZUNTS9VD56.

For citation purposes, cite each article independently as indicated on the article page online and as indicated below:

Lastname, A.A.; Lastname, B.B. Article Title. *Journal Name* **Year**, *Volume Number*, Page Range.

ISBN 978-3-7258-5149-2 (Hbk)
ISBN 978-3-7258-5150-8 (PDF)
https://doi.org/10.3390/books978-3-7258-5150-8

Contents

About the Editor

Yiquan Wu

Yiquan Wu is the Inamori Professor of Ceramics and Materials Science at the New York State College of Ceramics at Alfred University, USA. He is a Fellow of the American Ceramic Society (ACerS), an Academician of the World Academy of Ceramics, and has previously served as Chair of ACerS' Basic Science Division and President of the Ceramic Education Council. Additionally, he serves as an Editor for the Journal of the American Ceramic Society and as an Associate Editor for International Ceramic Engineering and Science. Dr. Wu received his Ph.D. in Materials from Imperial College London, England. His current research focuses on advanced ceramic materials. He has received numerous prestigious awards, including the NSF CAREER Award, the U.S. Air Force Office of Scientific Research (AFOSR) Young Investigator Award, the ACerS Global Ambassador Award, the Global Star Award from ACerS' Engineering Ceramics Division (ECD), the Empire Innovation Professor designation (State University of New York), and the Chancellor's Award for Excellence in Scholarship and Creative Activities from the State University of New York, among others.

ceramics

MDPI

Editorial

Recent Technological Advances in Transparent Ceramics

Yiquan Wu

Kazuo Inamori School of Engineering, New York State College of Ceramics, Alfred University, 2 Pine Street, Alfred, NY 14802-1296, USA; wuy@alfred.edu

1. Introduction

Transparent and translucent ceramics (TCs) represent a relatively recent development in the long history of ceramics—while silicate ceramics have existed for approximately 30,000 years, transparent ceramics have been developed only within the past 65 years. Together with certain glasses and single crystals, they form a class of materials known as transparent inorganic solids (TISs). A wide array of components and devices, including many high-tech applications, are derived from TISs [1–5]. The origins of the TC field can be traced to advancements in sintering science pioneered by R. L. Coble at General Electric in the late 1950s [1] (Figure 1). His work led to the development of translucent alumina walls for high-pressure sodium vapor lamps, which became a commercially successful product [6] and helped establish the field's significance.

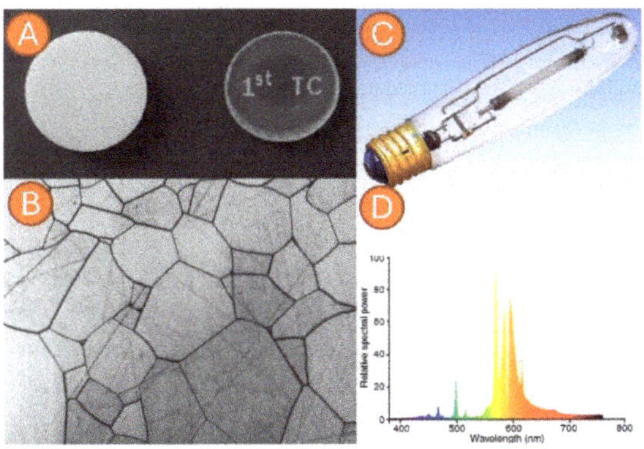

Figure 1. (**A**) First translucent alumina ceramics: regular alumina and translucent GE Lucalox alumina; (**B**) microstructure of translucent ceramics; (**C**) high-pressure sodium vapor lamp envelope made of translucent Lucalox alumina; (**D**) spectral composition of light emitted by the sodium vapor lamp. Reprinted with permission from [2]. Copyright 2020, John Wiley and Sons.

Many new materials and products have been developed within this evolving domain. For a long time, these materials were primarily based on cubic crystal lattices (unlike alumina), which significantly reduce light scattering and consequently enhance optical transmission. Notable examples include MgO [7,8], MgF_2 [9,10], perovskite-structured PLZT [11,12], ALON [13–15], $MgAl_2O_4$ [16,17], diamond [18], BN [19,20], and SiC [21]. These materials have enabled the development of primarily passive components, whose functionality depends on high optical transparency ($T > 70\%$) combined with desirable mechanical, thermal, electrical, or magnetic properties, alongside excellent chemical stability. Subsequently, more sophisticated materials were developed, in which transparent ceramics

 https://doi.org/10.3390/ceramics8030098

serve as hosts for intentional doping with transition-metal (TM) and/or rare-earth (RE) cations. These doped materials, often referred to as active transparent ceramics, can modify the spectral characteristics of incident light. Notable examples include $MgAl_2O_4$ [22], $ZnAl_2O_4$ [23,24], CaF_2 [25] and other fluorides [26], YAG [27–29], $Gd_3Ga_5O_{12}$ [30] and other garnets, cubic ZrO_2 [31], $Ca_5(PO_4)_3F$ [32], Y_2O_3 [33] and other bixbyite-structured sesquioxides [34], as well as oxide and FOX-type glass-ceramics (e.g., $BaAl_4O_7$) [35].

Passive transparent ceramics have enabled products such as transparent armor, infrared and LIDAR windows, cell phone displays, metal halide lamps, jewelry, and flash goggles. Active transparent ceramics, on the other hand, have led to the creation of solid-state lasers, optical filters, passive Q-switch absorbers, scintillators, and solid-state phosphors. A key objective for the near future is to reduce fabrication costs—particularly for spinel-based products. However, only modest progress is expected through incremental improvements to existing processes. Significant cost reductions are likely to be achieved only through the implementation of innovative, science-driven technologies. New materials are also expected to emerge, especially in the field of nanostructured, non-cubic transparent ceramics. Performance enhancements of current devices are anticipated in the short to medium term, alongside the discovery of new applications. A fundamental requirement for these advancements is the deepening and broadening of scientific knowledge underpinning the entire field.

2. Dr. Adrian Goldstein Author of Some Pioneering Contributions to the Transparent Ceramics Field

This Special Issue honors Dr. Adrian Goldstein, former head of the Israel Ceramics and Silicates Institute, in recognition of his groundbreaking contributions to the field of transparent ceramics. Dr. Goldstein was among the first to demonstrate that significant enhancement in densification rate could be achieved by microwave sintering of green bodies composed of monodisperse, spheroidal particles, such as amorphous silica, arranged in a highly compact configuration. His work showed that such accelerated densification enables the formation of transparent bulk specimens without inducing crystallization during the process [36]. Using a similar approach, Dr. Goldstein successfully fabricated transparent spinel with submicron grain sizes via low-temperature sintering of monodisperse, spheroidal powders at approximately 1300 °C—several hundred degrees lower than the temperatures typically required for conventional densification [37]. Extending these innovations, Dr. Goldstein also developed a novel method for producing bulk, polycrystalline, transparent $ZnAl_2O_4$ ceramics, introducing a new class of optical ceramics with a high hardness of 10.5 GPa [38].

Beyond fabrication, Dr. Goldstein conducted fundamental studies on the oxidation and coordination behavior of transition metal dopants in inverted ultrabasic silicate glasses. These investigations led to key discoveries, including the remarkable stability of Co^{3+} in ultra-high-basicity glass and a redox interaction mechanism between Cr^{3+} and Cu^{2+} ions [39]. Further advancing optical device technologies, Dr. Goldstein contributed to the development of Co^{2+}:$MgAl_2O_4$ transparent ceramics as saturable absorbers for passive Q-switching of Er^{3+} lasers at 1.534 μm. This work established Co^{2+}-doped $MgAl_2O_4$ ceramics as effective passive Q-switching elements [40]. In collaborative research, Dr. Goldstein also investigated the influence of cation inversion on the optical absorption behavior of transition-metal-doped spinel ceramics, notably showing that increased inversion levels in Ti-doped $MgGa_2O_4$ significantly alter its optical absorption spectrum, deepening our understanding of the structure–property relationships in spinel-type lattices [41]. Through these seminal contributions, Dr. Goldstein has profoundly impacted both the fundamental science and practical applications of transparent ceramics. His work continues to shape the future of ceramic materials research.

Dr. Adrian Goldstein, together with Dr. Andreas Krell and Prof. Zeev Burshtein, co-authored the excellent and comprehensive book Transparent Ceramics: Materials, Engineering and Applications, published by John Wiley & Sons, New York, in 2020.

The book *Transparent Ceramics: Materials, Engineering and Applications*, represents a long-awaited and much-needed comprehensive overview of transparent ceramics. From the outset, the book clearly communicates the breadth and depth of its content. Its accessible and engaging writing style avoids overly technical language, making it suitable not only for students, scientists, and engineers in materials science but also for newcomers to the field. The material is richly supported by numerous tables, diagrams, and illustrations, enhancing the reader's experience and facilitating comprehension. While significant progress has been made in fabricating high-quality transparent ceramics, many challenges and open questions remain. These are thoughtfully addressed in the final chapter, Future Developments, which outlines the author's perspective on the field's future directions. This book is an invaluable resource and holds great potential to become the definitive textbook for anyone interested in transparent ceramics.

3. Works Collected in This Special Issue

Recognizing the growing importance of transparent ceramics, the MDPI journal *Ceramics* is committed to publishing research focused on their science and engineering. This Special Issue brings together ten such contributions. A brief overview of their content is provided below.

The first contribution [42] examines the effect of sintering atmosphere on the densification behavior of Yb-doped Y_2O_3 transparent ceramics, suited for high-power laser applications. It discusses the potential advantages of flowing oxygen versus traditional vacuum environments during the sintering process. The second paper [43] presents a large-scale commercial application of a sophisticated transparent ceramic coating designed to protect rose wines from degradation caused by UV and blue light exposure. The coatings were prepared from a sol–gel precursor capable of forming silica-based gels. Commercial organic UV-absorbing molecules were incorporated into the gel, and two different organic additives were evaluated. The resulting coatings were colorless and thin, demonstrating superior UV protection compared to gels with inorganic additives. The most effective formulation, containing approximately 1.5% of the commercial additive SemaSORB 20109, exhibited the highest efficacy in preventing anthocyanin loss. The next study [44] also focuses on high-power lasers, investigating the fabrication of Yb-doped components using a CaF_2 host. Despite its relatively low mechanical strength, CaF_2 offers several advantages over oxide hosts, such as its low phonon energy. Moreover, CaF_2-based transparent ceramics can be sintered at temperatures below 1000 °C, whereas oxide ceramics typically require sintering temperatures near 2000 °C. The authors employed a two-step densification involving pressureless sintering in air followed by hot isostatic pressing (HIP) at 600 °C. The resulting specimens achieved approximately 91% optical transmission at 1.2 μm. Under quasi-continuous wave (QCW) operation at ~1030 nm, with an output coupler mirror transmission of 18.1%, the laser delivered a maximum output power of 1.51 W and a slope efficiency of 9.2%.

The study presented in [45] was a collaborative effort among researchers from Algeria, Spain, and France. It addressed the restoration of sand-blasted Mg-spinel ceramic specimens—fabricated via spark plasma sintering (SPS)—by applying a thin, high-hardness SiO_2–ZrO_2 protective layer. Sand blasting simulated natural weathering effects in desert environments. The protective coating was synthesized and deposited via sol–gel processing. Interestingly, sand blasting was found to improve UV-region transmission (around 200 nm) in specimens with initially low transparency, while degrading transmission in highly transparent samples. Larger defects from blasting mainly reduced transmission and increased surface roughness at longer wavelengths. The applied coating exhibited notable healing capabilities, successfully restoring the original transmission characteristics. Paper [46] explores the development of transparent ceramics (TCs) from non-cubic materials, specifically $Ca_{10}(PO_4)_6(OH)_2$, a phosphate-based hydroxyapatite. Such materials have already demonstrated potential as solid-state laser hosts. The study focused on optimizing processing conditions using the spark plasma sintering (SPS) technique. A comparison was made between commercial nano-powders and in-house synthesized powders via hydrothermal methods. Initial tests led to the successful fabrication of 1 mm thick disks with approximately 60% optical transmission at 880 nm.

The following contribution [47] addresses an important characterization method: scanning optical microscopy. This technique enables a detailed examination of porosity and other defects in transparent ceramics. Modern digital instruments enable the acquisition of high-resolution images across multiple depths and relatively large surface areas. Although small pores are better revealed through electron microscopy, optical scanning provides valuable data on pore volume, size, and shape distribution. Yb-doped YAG ceramics were analyzed, with 2.2×10^7 μm^3 examined per sample. The study revealed clear differences in porosity among specimens fabricated using different techniques. Samples that appeared similar under visual inspection or SEM exhibited distinct pore populations, with these differences correlating well with their respective transmission spectroscopy data.

Paper [48] focuses on the development of eye-safe lasers, specifically examining Ho^{3+}-doped Lu_2O_3 transparent ceramics fabricated via spark plasma sintering (SPS). The goal was to compare the properties of these SPS-fabricated ceramics with those of single crystals and ceramics densified by hot isostatic pressing (HIP). Using in-house-prepared active powders, specimens with characteristics comparable to single crystals were successfully produced. However, as the Ho^{3+} concentration approached 10%, auto-quenching became a concern. Energy losses were attributed to trace impurities, but cross-relaxation processes enhanced emission near 2 μm. Paper [49] investigates the challenge of joining transparent Mg-spinel plates. It is now well understood that although fabricating large spinel windows is technically possible, it remains economically impractical. Consequently, joining smaller plates appears to be the most viable approach for producing large, transparent armor components. The study explored bonding components using specially designed glass interlayers, aiming to develop a joining glass with refractive index and thermal expansion closely matched to spinel. Additionally, the glass should minimize spinel dissolution during thermal treatment and resist both crystallization and environmental degradation. A particularly interesting and practically significant phenomenon was observed when two semi-polished spinel plates were bonded using this glass. In this context, "semi-polished" refers to one plate face being well polished and the other subjected only to coarse grinding, with the glass applied to the ground face. Remarkably, the resulting bonded specimens exhibited optical transmission exceeding the theoretical value for fully polished, uncoated plates—representing a highly valuable achievement in the field.

Paper [50] details the fabrication of transparent ceramic spinel (TCS) that combines high optical transparency with excellent resistance to sand erosion—primarily achieved through high hardness. The study identifies spark plasma sintering (SPS) as a powerful method for producing components with the desired properties. Importantly, it was found that the highest sintering temperature did not yield optimal results. Instead, 1350 °C was identified as the best compromise, balancing competing factors such as pore coalescence, grain growth, graphite contamination, and diffusion kinetics. The resulting transparent samples achieved Vickers hardness of 16 GPa and Young's modulus of 270 MPa, surpassing the properties of parts densified by hot isostatic pressing (HIP). The subsequent paper [51] examines the speciation of transition metals (TMs) incorporated into transparent ceramics. Understanding speciation—together with insights from ligand field theory—is essential for interpreting optical spectra and predicting spectral line profiles. This study investigated how differences in host lattice structures affect TM speciation, using corundum and spinel as model systems. It was observed that the spinel structure favors the stabilization of TMs in the 2+ oxidation state (substituting Mg^{2+} in tetrahedral sites), while corundum tends to favor 3+ cations occupying octahedral sites. Additionally, variation in ionization potential along the 3d transition series contributes to the stabilization of lower oxidation states for late transition metals. Normal crystallographic sites were found to be significantly more favorable for TM incorporation than inverted ones.

Acknowledgments: The author appreciates Adrian Goldstein very much for his support and comments on this Special Issue and Editorial Article.

Conflicts of Interest: The author declares no conflict of interest.

References

1. Coble, R.L. Sintering of Alumina: Effect of Atmospheres. *J. Amer. Ceram. Soc.* **1962**, *54*, 123–127. [CrossRef]
2. Goldstein, A.; Krell, A.; Burshtein, Z. *Transparent Ceramics: Materials, Engineering and Applications*; Wiley: New York, NY, USA, 2020.
3. Goldstein, A.; Krell, A. Transparent Ceramics at 50: Progress made and Further Prospects. *J. Am. Ceram. Soc.* **2016**, *99*, 3173–3197. [CrossRef]
4. Ikesue, A.; Aung, Y.L.; Lupei, V. *Ceramic Lasers*; CUP: Cambridge, UK, 2013.
5. Kong, L.B.; Huang, Y.Z.; Que, W.X.; Zhang, T.S.; Li, S.; Zhang, J.; Dong, Z.L.; Tang, D.Y. *Transparent Ceramics*; Springer: Heidelberg, Germany, 2015.
6. Burke, J.E. Lucalox alumina: The ceramics which revolutionized outdoor lighting. *MRS Bull.* **1996**, *21*, 61–68. [CrossRef]
7. Chen, X.; Wu, Y. Fabrication and optical properties of highly transparent MgO ceramics by spark plasma sintering. *Scr. Mater.* **2019**, *162*, 14–17. [CrossRef]
8. Chen, X.; Zhang, G.; Tomala, R.; Hreniak, D.; Wu, Y. Yb doped MgO transparent ceramics generated through the SPS method. *J. Eur. Ceram. Soc.* **2022**, *42*, 4320–4327. [CrossRef]
9. Li, X.; Yin, M.; Xiao, C.; Kuang, Z.; Yu, S.; Li, X.; Jia, Z. Preparation of highly transparent mid-infrared MgF_2 ceramics by low-pressure hot pressing. *Opt. Mater.* **2024**, *157*, 116286. [CrossRef]
10. Zavala-Cuéllar, N.; Gómez-Solís, C.; Vallejo, M.A.; Gómez, M.R.; Cerón, P.; León-Madrid, M.I. Study of the effect of the precursor salt on the synthesis of magnesium fluoride and its influence on the optical properties. *Ceram. Int.* **2024**, *50*, 24939–24947. [CrossRef]
11. Milisavljevic, I.; Zhang, M.; Jiang, Q.; Liu, Q.; Wu, Y. Transparent electro-optic ceramics: Processing, materials, and applications. *J. Mater.* **2025**, *11*, 100872. [CrossRef]
12. Chen, X.; Chen, R.; Chen, Z.; Chen, J.; Shung, K.K.; Zhou, Q. Transparent lead lanthanum zirconate titanate (PLZT) ceramic fibers for high-frequency ultrasonic transducer applications. *Ceram. Int.* **2016**, *42*, 18554–18559. [CrossRef]
13. Zhu, D.; Zhou, J.; Zheng, J.; Huo, T.; Dai, Y.; Wu, J. Preparation of transparent AlON ceramics with controlled shapes by a novel spontaneous coagulation casting and pressureless sintering method. *Ceram. Int.* **2024**, *50*, 22373–22380. [CrossRef]
14. Yang, S.; Li, J.; Guo, H.; Mao, X.; Tian, R.; Zhang, J.; Wang, S. Reactive sintered highly transparent AlON ceramics with Y_2O_3-$MgAl_2O_4$-H_3BO_3 ternary additive. *J. Am. Ceram. Soc.* **2021**, *104*, 4304–4308. [CrossRef]
15. Yang, J.; Zhang, H.; Li, Y.; Xu, W.; Zhou, Y. Maochun Hong AlON transparent ceramics from powders synthesized by improved direct nitridation. *Ceram. Int.* **2023**, *49*, 35991–36001. [CrossRef]
16. Merac, M.R.D.; Kleebe, H.-J.; Müller, M.M.; Reimanis, I.E. Fifty Years of Research and Development Coming to Fruition; Unraveling the Complex Interactions during Processing of Transparent Magnesium Aluminate ($MgAl_2O_4$) Spinel. *J. Am. Ceram. Soc.* **2013**, *96*, 3341–3365. [CrossRef]
17. Cottrino, S.; Gaudisson, T.; Douillard, T.; Blanchard, N.; Meille, S.; Gremillard, L.; Mercury, M.; Le Floch, S. Effect of high pressure on microstructure and mechanical and optical properties of nano-structured $MgAl_2O_4$ spinel fabricated by High Pressure Spark Plasma Sintering. *J. Eur. Ceram. Soc.* **2025**, *45*, 117579. [CrossRef]

18. Zhang, J.; Zhan, G.; He, D.; Li, D.; Li, Q.; Du, C.; Da, Q.; Liu, F.; Yan, X. Transparent diamond ceramics from diamond powder. *J. Eur. Ceram. Soc.* **2023**, *43*, 853–861. [CrossRef]
19. Zhao, M.; Kou, Z.; Zhang, Y.; Peng, B.; Wang, Y.; Wang, Z.; Yin, X.; Jiang, M.; Guan, S.; Zhang, J.; et al. Superhard transparent polycrystalline cubic boron Nitride. *Appl. Phys. Lett.* **2021**, *118*, 151901–151906. [CrossRef]
20. Taniguchi, T.; Akaishi, M.; Yamaoka, S. Mechanical Properties of Polycrystalline Translucent Cubic Boron Nitride as Characterized by the Vickers Indentation Method. *J. Am. Ceram. Soc.* **1996**, *79*, 547–549. [CrossRef]
21. Bayya, S.S.; Villalobos, G.R.; Hunt, M.P.; Sanghera, J.S.; Sadowski, B.; Aggarwal, I.D.; Cinibulk, M.; Carney, C.; Keller, K. Development of Transparent Polycrystalline Beta-Silicon Carbide. In Proceedings of the SPIE Optical Engineering + Applications 2013, San Diego, CA, USA, 25–29 August 2013; Volume 8837, p. 88370S-1.
22. Wang, X.; Lu, T.; Sun, Y. Highly doped Mn:$MgAl_2O_4$ Transparent Ceramics. *Key Eng. Mater.* **2008**, *368–372*, 414–416. [CrossRef]
23. Cornu, L.; Gaudon, M. Veronique Jubera $ZnAl_2O_4$ as potential sensor: Variation of luminescence with thermal history. *J. Mater. Chem. C* **2013**, *1*, 5419–5428. [CrossRef]
24. Kim, B.-N.; Hiraga, K.; Jeong, A.; Hu, C.; Suzuki, T.S.; Yun, J.-D.; Sakka, Y. Transparent $ZnAl_2O_4$ ceramics fabricated by spark plasma sintering. *J. Ceram. Soc. Jpn.* **2014**, *122*, 784–787. [CrossRef]
25. Uehara, H.; Yao, W.; Ikesue, A.; Noto, H.; Chen, H.; Hishinuma, Y.; Muroga, T.; Yasuhara, R. Dy-doped CaF_2 transparent ceramics as a functional medium in the broadband mid-infrared spectral region. *Opt. Contin.* **2020**, *3*, 1811–1818. [CrossRef]
26. Yu, S.; Wu, Y. Fabrication, microstructure and optical properties of Ce:SrF_2 transparent ceramics. *Opt. Mater.* **2020**, *105*, 109898. [CrossRef]
27. Zhang, G.; Carloni, D.; Wu, Y. Ultraviolet emission transparent Gd:YAG ceramics processed by solid-state reaction spark plasma sintering. *J. Am. Ceram. Soc.* **2020**, *103*, 839–848. [CrossRef]
28. Chen, X.; Wu, Y.; Wei, N.; Qi, J.; Lu, Z.; Zhang, Q.; Hua, T.; Zeng, Q.; Lu, T. The roles of cation additives on the color center and optical properties of Yb:YAG transparent ceramic. *J. Eur. Ceram. Soc.* **2018**, *38*, 1957–1965. [CrossRef]
29. Chen, S.; Zhang, L.; Kisslinger, K.; Wu, Y. Transparent $Li_{0.05}Y_3Al_5O_{12}$:$Ce^{3+}{}_{0.01}$ Ceramics for Thermal Neutron Detection. *J. Am. Ceram. Soc.* **2013**, *96*, 1067–1069. [CrossRef]
30. Liu, M.; Zhang, Y.; Hu, S.; Zhou, G.; Qin, X.; Wang, S. Preparation of Ce-Doped $Gd_3(Al, Ga)_5O_{12}$ Nanopowders via Microwave-Assisted Homogenization Precipitation for Transparent Ceramic Scintillators. *Materials* **2024**, *17*, 1258. [CrossRef]
31. Kalaivani, S.; Ezhilan, M.; Deepa, M.; Kannan, S. Rare earth doped Zirconia: Structure, physicochemical properties and recent advancements in technological applications. *Prog. Solid State Chem.* **2025**, *79*, 100524. [CrossRef]
32. Furuse, H.; Kato, D.; Morita, K.; Suzuki, T.S.; Kim, B.-N. Characterization of Transparent Fluorapatite Ceramics Fabricated by Spark Plasma Sintering. *Materials* **2022**, *15*, 8157. [CrossRef] [PubMed]
33. Stanciu, G.; Gheorghe, L.; Voicu, F.; Hau, S.; Gheorghe, C.; Croitoru, G.; Enculescu, M.; Yavetskiy, R.P. Highly transparent Yb:Y_2O_3 ceramics obtained by solid-state reaction and combined sintering procedures. *Ceram. Int.* **2019**, *45*, 3217–3222. [CrossRef]
34. Yin, D.; Ma, J.; Liu, P.; Yao, B.; Wang, J.; Dong, Z.; Kong, L.B.; Tang, D. Submicron-grained Yb:Lu_2O_3 transparent ceramics with lasing quality. *J. Am. Ceram. Soc.* **2019**, *102*, 2587–2592. [CrossRef]
35. Allix, M.; Alahrache, S.; Fayon, F.; Suchomel, M.; Porcher, F.; Cardinal, T.; Matzen, G. Highly transparent $BaAl_4O_7$ polycrystalline ceramic obtained by full crystallization from glass. *Adv. Mater.* **2012**, *24*, 5570–5575. [CrossRef] [PubMed]
36. Goldstein, A.; Ruginets, R.; Geffen, Y. Microwave sintering of amorphous silica powders. *J. Mater. Sci. Lett.* **1997**, *16*, 310–312. [CrossRef]
37. Goldstein, A.; Goldenberg, A.; Hefet, M. Transparent polycrystalline $MgAl_2O_4$ spinel with submicron grains by low temperature sintering. *J. Ceram. Soc. Jpn.* **2009**, *117*, 1281–1283. [CrossRef]
38. Goldstein, A.; Yeshurun, Y.; Vulfson, M.; Kravits, H. Fabrication of Transparent Polycrystalline $ZnAl_2O_4$—A New Optical Bulk Ceramic. *J. Am. Ceram. Soc.* **2012**, *95*, 879–882. [CrossRef]
39. Zandona, A.; Castaing, V.; Shames, A.I.; Helsch, G.; Deubener, J.; Becerro, A.I.; Allix, M.; Goldstein, A. Oxidation and coordination states assumed by transition metal dopants in an invert ultrabasic silicate glass. *J. Non-Cryst. Solids* **2023**, *603*, 122094. [CrossRef]
40. Shirakov, A.; Burshtein, Z.; Goldstein, A.; Frumker, E.; Ishaaya, A. Use of Co^{2+}:$MgAl_2O_4$ transparent ceramics in passive Q-switching of an Er:glass laser at 1.53 μm. *Opt. Express* **2020**, *28*, 21956–21970. [CrossRef]
41. Zhang, G.; Wu, Y.; Shemes, A.; Goldstein, A. Influence of inversion level on the optical absorption spectra of Ti-doped transparent $MgGa_2O_4$ ceramics. *J. Am. Ceram. Soc.* **2022**, *105*, 5944–5955. [CrossRef]
42. Wang, X.; Xing, D.; Wang, Y.; Wang, J.; Ma, J.; Liu, P.; Zhang, J.; Tang, D. Optimizing Sintering Conditions for Y_2O_3 Ceramics: A Study of Atmosphere-Dependent Microstructural Evolution and Optical Performance. *Ceramics* **2025**, *8*, 66. [CrossRef]
43. Moriones, J.; Osés, J.; Amézqueta, P.; Palacio, J.F.; De Ara, J.F.; Almandoz, E. Enhancement of Sol–Gel Coatings for Photoprotection of Rosé Wines. *Ceramics* **2025**, *8*, 17. [CrossRef]
44. Li, X.; Hu, C.; Guo, L.; Wu, J.; Toci, G.; Pirri, A.; Patrizi, B.; Vannini, M.; Liu, Q.; Hreniak, D.; et al. Optimization of Yb:CaF_2 Transparent Ceramics by Air Pre-Sintering and Hot Isostatic Pressing. *Ceramics* **2024**, *7*, 1053–1065. [CrossRef]

45. Zegadi, A.; Ayadi, A.; Khellaf, I.; Hamidouche, M.; Fantozzi, G.; Durán, A.; Castro, Y. Improving the Transparency of a $MgAl_2O_4$ Spinel Damaged by Sandblasting through a SiO_2-ZrO_2 Coating. *Ceramics* **2024**, *7*, 743–758. [CrossRef]
46. Prokop, K.A.; Cottrino, S.; Garnier, V.; Fantozzi, G.; Guyot, Y.; Boulon, G.; Guzik, M. Enhancing Transparency in Non-Cubic Calcium Phosphate Ceramics: Effect of Starting Powder, LiF Doping, and Spark Plasma Sintering Parameters. *Ceramics* **2024**, *7*, 607–624. [CrossRef]
47. Picelli, F.; Hostaša, J.; Piancastelli, A.; Biasini, V.; Melandri, C.; Esposito, L. Beyond Scanning Electron Microscopy: Comprehensive Pore Analysis in Transparent Ceramics Using Optical Microscopy. *Ceramics* **2024**, *7*, 401–410. [CrossRef]
48. Viers, L.; Guené-Girard, S.; Dalla-Barba, G.; Jubéra, V.; Cormier, É.; Boulesteix, R.; Maître, A. Optical and Spectroscopic Properties of Ho:Lu_2O_3 Transparent Ceramics Elaborated by Spark Plasma Sintering. *Ceramics* **2024**, *7*, 208–221. [CrossRef]
49. Hormadaly, J. Glass Composition for Coating and Bonding of Polycrystalline Spinel Ceramic Substrates. *Ceramics* **2024**, *7*, 101–114. [CrossRef]
50. Hoggas, K.; Benaissa, S.; Cherouana, A.; Bouheroum, S.; Assali, A.; Hamidouche, M.; Fantozzi, G. Mechanical Behavior of Transparent Spinel Fabricated by Spark Plasma Sintering. *Ceramics* **2023**, *6*, 1191–1209. [CrossRef]
51. Goldstein, A.; Zandonà, A. Speciation of 3d Elements in Spinel Versus Corundum: Elucidating the Interplay Between Ligand Field, Structural Dissimilarities and Processing Conditions. *Ceramics* **2025**, *8*, 16. [CrossRef]

ceramics

MDPI

Article

Optimizing Sintering Conditions for Y_2O_3 Ceramics: A Study of Atmosphere-Dependent Microstructural Evolution and Optical Performance

Xueer Wang [1,†], Dongliang Xing [1,†], Ying Wang [1,*], Jun Wang [1], Jie Ma [1], Peng Liu [1], Jian Zhang [2] and Dingyuan Tang [3,*]

1 Jiangsu Key Laboratory of Advanced Laser Materials and Devices, School of Physics and Electronic Engineering, Jiangsu Normal University, Xuzhou 221116, China; 17625047021@163.com (X.W.); 19351711055@163.com (D.X.); jwang025@e.ntu.edu.sg (J.W.); jiema07@gmail.com (J.M.); liupeng_tju@126.com (P.L.)
2 Shanghai Institute of Ceramics, Chinese Academy of Sciences, Shanghai 201899, China; jianzhang@mail.sic.ac.cn
3 Future Technology School, Shenzhen Technology University, Shenzhen 518118, China
* Correspondence: ywang31@e.ntu.edu.sg (Y.W.); tangdingyuan@sztu.edu.cn (D.T.)
† These authors contributed equally to this work.

Abstract: This paper systematically investigated the influence of sintering atmospheres, vacuum, and oxygen, on the microstructure and optical properties of Y_2O_3 ceramics. Compared with vacuum sintering, sintering in flowing oxygen atmosphere can effectively inhibit the grain growth of Y_2O_3 ceramics at the final stage of sintering and improve the uniformity of microstructure. After hot isostatic pressing, the samples pre-sintered at oxygen atmosphere showed good in-line transmittance from a visible-to-mid-infrared wavelength range (0.4–6.0 μm) except in the range of 2.8–4.1 μm. Spectral analysis showed that an obvious broadband absorption peak (2.8–4.1 μm) of characteristic hydroxyl groups is detected in the above samples. However, before densification, a low-temperature heat treatment at 600 °C under vacuum can effectively diminish the hydroxyl groups in Y_2O_3 ceramics. However, laser experiments in the ~1 μm wavelength range showed that although the Yb:Y_2O_3 ceramic carrying hydroxyl had obvious absorption in the 2.8–4.1 μm range, it had little effect on its laser oscillation in the ~1 μm wavelength. Yb:Y_2O_3 ceramics pre-sintered in an oxygen atmosphere at 1460 °C followed by hot isostatic pressing at 1440 °C achieved 12.85 W continuous laser output at room temperature, with a laser slope efficiency of 84.4%.

Keywords: Y_2O_3 ceramics; oxygen atmosphere sintering; hydroxyl group; broadband absorption

1. Introduction

In recent decades, yttrium oxide (Y_2O_3) ceramics have garnered significant attention due to their superior properties, including low phonon energy, high thermal conductivity, good thermal stability, and broad optical transparency (0.22–8 μm) [1–3]. Consequently, Y_2O_3 transparent ceramics are considered ideal candidates for applications in high-power laser systems, high-temperature lens, and infrared windows [4,5]. As a cost-effective alternative to obtain Y_2O_3 materials with inherent high melting temperature (>2400 °C), the ceramic counterparts can be fabricated under much lower sintering temperature (<1800 °C), meanwhile demonstrating the advantages of composite structure design and volume scalability [6,7]. However, achieving high-quality Y_2O_3 ceramics remains challenging, with demands to suppress residual pores, exaggerated grain growth, and impurity incorporation (e.g., hydroxyl groups) that degrade optical transparency. Some defects are closely related to the sintering conditions, particularly sintering atmosphere and temperature, that can greatly influence the microstructure evolution and densification process [8,9].

Conventional sintering methods for Y_2O_3 ceramics typically involve vacuum sintering [10], which can avoid the entrapment of insoluble gas within pores, thereby promoting final full densification. However, during the final-stage sintering of Y_2O_3 ceramics, accelerated grain boundary mobility can lead to pore–boundary separation, resulting in the formation of intragranular pores that are difficult to remove, even further with pressure-assisted hot isostatic pressing (HIP) sintering [11]. Therefore, it is necessary to control microstructure evolution during the vacuum sintering process. Additionally, it is well known that the grain boundary mobility of Y_2O_3 ceramics is rate-controlled by yttrium cation interstitial concentrations, which is inversely proportional to the concentration of oxygen interstitials [9,12]. Under an oxygen-rich sintering atmosphere, the concentration of oxygen interstitials in Y_2O_3 tends to increase, thus inhibiting the abnormal grain growth. Therefore, sintering in an oxygen atmosphere can serve as an effective way to fabricate highly dense Y_2O_3 transparent ceramics.

In this paper, a systematic comparison was conducted on the effects of vacuum and oxygen pre-sintering on the microstructure and densification behaviors of Y_2O_3 ceramics. The optical properties, including transmittance and absorption spectra, were compared with different sintering methods. A laser experiment was also conducted on Yb:Y_2O_3 ceramics, which were pre-sintered under oxygen atmosphere.

2. Experimental Section

2.1. Ceramic Fabrication

Y_2O_3 and 3 at.% Yb:Y_2O_3 transparent ceramics were prepared by the chemical co-precipitation method. Firstly, commercial high-purity Y_2O_3 and Yb_2O_3 powder was used as the raw material. They were dissolved in dilute hydrochloric acid solution. According to the chemical formula of Y_2O_3 and $(Yb_{0.03}Y_{0.97})_2O_3$, the prepared YCl_3 and $YbCl_3$ solutions were mixed to make the concentration of Y^{3+} reach 0.2 mol/L. Next, a mixture of NH_4OH and NH_4HCO_3 was used as the precipitant (with molar concentrations of 1.5 mol/L and 2 mol/L, respectively). The precipitant was sprayed into the mother solutions at a rate of 10 mL/min until the pH value reached 8. The obtained precursors were aged at 15 °C for about 5 h, then filtered and washed 4 times with ultrapure water and 3 times with ethanol to remove reaction by-products. Subsequently, the washed precursors were dried in an oven at 80 °C for about 48 h. Then, they were crushed and calcined in a muffle furnace at 1200 °C/5 h. Then, the obtained Y_2O_3 and Yb:Y_2O_3 powders were ball-milled by using a planetary milling machine. Ethanol and ZrO_2 balls were used as the ball-milling medium. The solid contents of the slurries were fixed at 11 vol%. The rotational speed of the machine was set to 140 r/min, with a milling duration of 15 h. The resulting slurries were completely dried and sieved through a 100-mesh sieve. The sieved powder was then calcined at 700 °C for 10 h to remove any by-products. The calcined powder was weighed and poured into a mold with a diameter of 22 mm. A uniaxial pressure of 5 MPa was applied and maintained for 5 min. The resulting green compact, formed via dry pressing, was subsequently subjected to cold isostatic pressing at 200 MPa to obtain ceramic green bodies. The green bodies underwent pre-sintering in either oxygen (oxygen purity 99.9%) or vacuum (vacuum degree $< 5 \times 10^{-3}$ Pa) atmospheres. The pre-sintering temperature and duration were 1400–1550 °C and 7 h, respectively. Subsequently, some of the samples were further densified via hot isostatic pressing (HIP) at 198 MPa in an argon atmosphere for 3 h at 1440 °C. After HIP treatment, the samples were annealed in air using a muffle furnace at 1000 °C for 10 h. Finally, the transparent ceramic samples were double-sided polished.

2.2. Characterization

The relative densities of the samples were tested using Archimedes' method. The microstructures of the sintered ceramics were observed using a scanning electron microscope (SEM, JSM-6510, JEOL, Kariya, Japan). Before observation, the samples were mirror polished on the surface and then thermally etched at 1250 °C for 5 h in a muffle furnace. In-line transmittances in the visible to near-infrared wavelength range were measured using a UV–Vis-NIR spectrophotometer (Lambda 950, Perkin-Elmer, Waltham, MA, USA). The measurement wavelength range was from 200 to 2500 nm with a step size of 1 nm. In-line transmittances in the mid-infrared range were measured using a Fourier-transform infrared spectrometer (FTIR, TENSOR 27, Bruker, Germany). All the measurements were carried out at room temperature.

3. Results and Discussion

Figure 1 compares the evolution of microstructures in Y_2O_3 ceramics as a function of sintering temperature during the final sintering stage under flowing oxygen and vacuum conditions. Samples sintered in an oxygen atmosphere exhibit smaller average grain sizes and lower porosity compared to those sintered in vacuum at identical temperatures (Figure 1a–d,f–i). Notably, the oxygen-sintered specimens show significantly narrower grain size distributions (Figure 1e,j), which can be attributed to two primary factors. First, the oxygen-rich environment promotes the formation of oxygen interstitials within the crystal lattice, effectively reducing the grain boundary mobility of Y_2O_3 and thereby suppressing grain growth [12]. Second, the forced convection induced by the flowing gas atmosphere enhances heat transfer efficiency within the furnace chamber. In contrast, vacuum sintering primarily relies on thermal radiation for heat transfer, leading to less efficient temperature distribution. The improved thermal uniformity in oxygen sintering ultimately contributes to a more homogeneous microstructure in the sintered specimens.

Figure 2a compares the evolution of the relative density of Y_2O_3 ceramics as a function of sintering temperature under different sintering atmospheres. When the sintering temperature is below 1430 °C, samples sintered in vacuum exhibit slightly higher relative density than those sintered in an oxygen atmosphere. This can be attributed to the increased concentration of oxygen vacancies in vacuum, which enhances grain boundary diffusion in Y_2O_3 ceramics, thereby promoting densification during the intermediate stages of sintering. However, during the final stage of sintering, the number of residual pores decreases significantly, leading to a substantial reduction in the pinning effect on grain boundaries. At this stage, the high grain boundary diffusion rate facilitates rapid grain growth, as shown in Figure 2b. As grain growth progresses, residual pores tend to coalesce and migrate along with the grain boundaries, resulting in a gradual decrease in pore number and an increase in pore size, as illustrated in Figure 3. With the continuous increase in pore size, their mobility declines, and they finally become trapped within the rapidly migrating grain boundaries, forming intragranular pores. As shown in Figures 2 and 3, Y_2O_3 ceramics sintered in an oxygen atmosphere exhibit smaller average grain sizes, lower porosity and smaller average pore sizes at the same sintering temperature compared to those sintered in vacuum. This can be attributed to enhanced microstructural uniformity and the suppression of excessive grain growth.

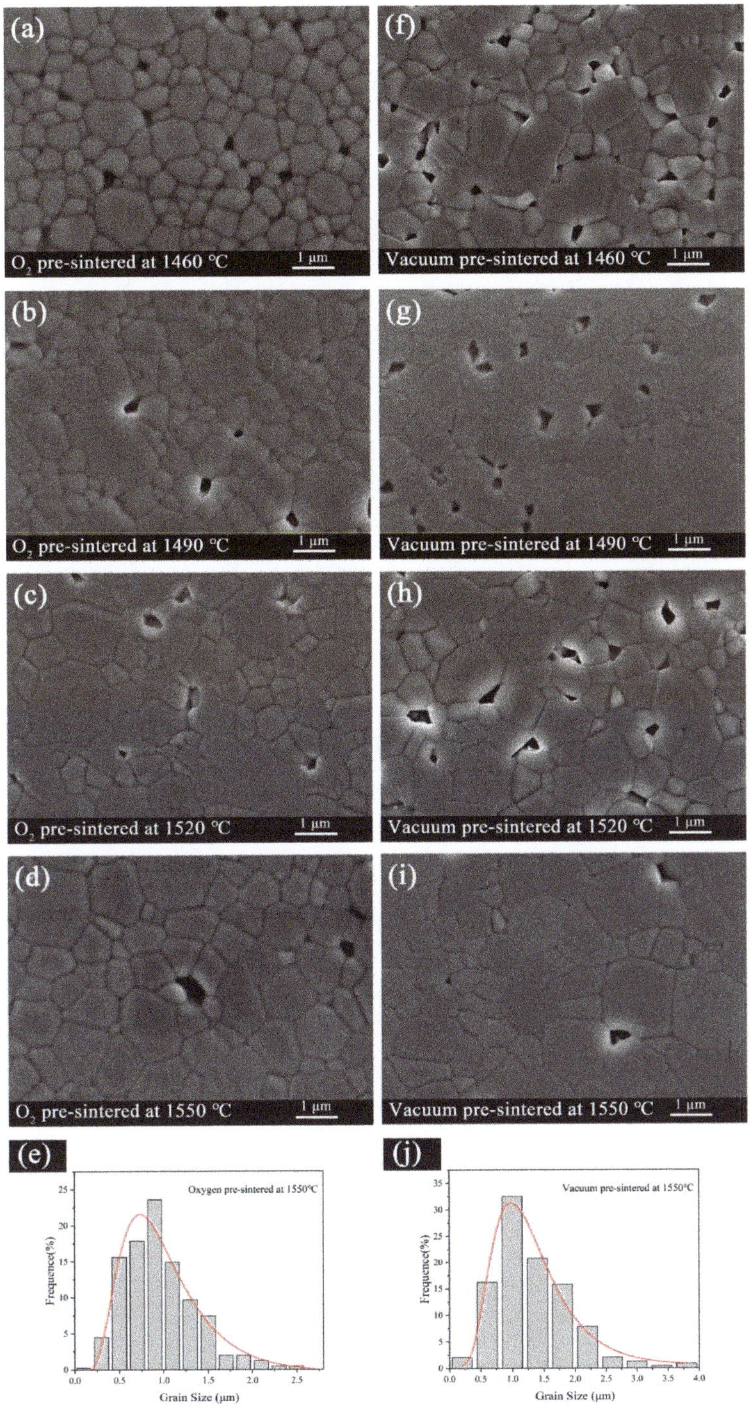

Figure 1. The comparisons of microstructures (**a–d**,**f–i**) and grain size distributions (**e**,**j**) of Y_2O_3 ceramics as a function of sintering temperature under flowing oxygen and vacuum conditions.

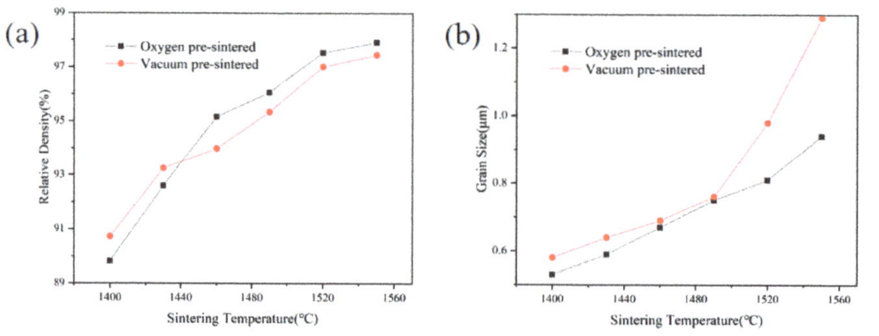

Figure 2. Comparisons of relative densities (**a**) and average grain sizes (**b**) of Y_2O_3 ceramics as a function of sintering temperatures under different sintering atmospheres.

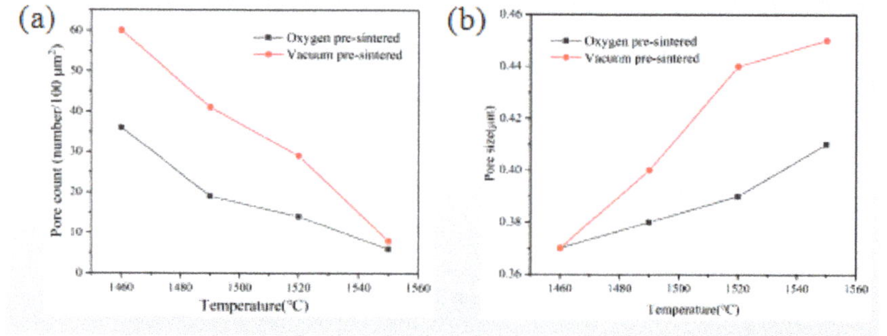

Figure 3. Variation in pore number density (**a**) and average pore size (**b**) with sintering temperature under different sintering atmospheres.

To further eliminate residual pores in Y_2O_3 ceramics, samples pre-sintered under O_2 and vacuum atmospheres were both subjected to hot isostatic pressing. Figure 4 compares the XRD of Y_2O_3 ceramics that underwent vacuum or oxygen pre-sintering at 1460 °C, followed by HIP at 1440 °C. As shown in Figure 4a, the XRD patterns of the two ceramics correspond to the diffraction peaks of the pure cubic Y_2O_3 phase (PDF no.74-1828), and no extra impurity peak is found, indicating that the sample does not produce a second phase. As shown in Figure 4b, compared with the vacuum pre-sintering sample, the X-ray diffraction peak of oxygen pre-sintering Y_2O_3 ceramics moves slightly to the small angle direction. This is because the introduction of oxygen interstitial point defects leads to an increase in the lattice constant of Y_2O_3, which shows that the X-ray diffraction peak of vacuum pre-sintering Y_2O_3 ceramics deviates to the small angle direction.

Figure 5 compares the transmittance of Y_2O_3 ceramics that underwent vacuum or oxygen pre-sintering at 1460 °C, followed by HIP at 1440 °C. Representative sample properties of the transparent Y_2O_3 ceramics in recent work are also listed in Table 1 for comparison. As shown in Figure 5a, Y_2O_3 ceramics pre-sintered under both atmospheres exhibit high in-line transmittance in the visible and near-infrared regions (the sample thickness is 4.3 mm). The transmittance exceeds 80% at wavelengths above 700 nm, indicating that the samples pre-sintered in either vacuum or oxygen exhibit good optical quality. However, as seen in Figure 5b, the sample pre-sintered in oxygen exhibits a broad absorption peak in the 2.8–4.1 μm range, which corresponds to the characteristic absorption of hydroxyl groups [13]. This phenomenon indicates that hydroxyl groups in Y_2O_3 ceramics can only be effectively removed through sintering in a vacuum or a reducing atmosphere.

Their resistance to removal at high temperatures under oxygen suggests that hydroxyl groups are not simply present in the form of absorbed water molecules within the Y_2O_3 green bodies but are chemically bonded to yttrium ions. During high-temperature sintering, the desorption reaction of hydroxyl groups from the material is shown in Reaction (1) [14].

$$OH^- + OH^- \leftrightarrow O^{2-} + H_2O \uparrow \tag{1}$$

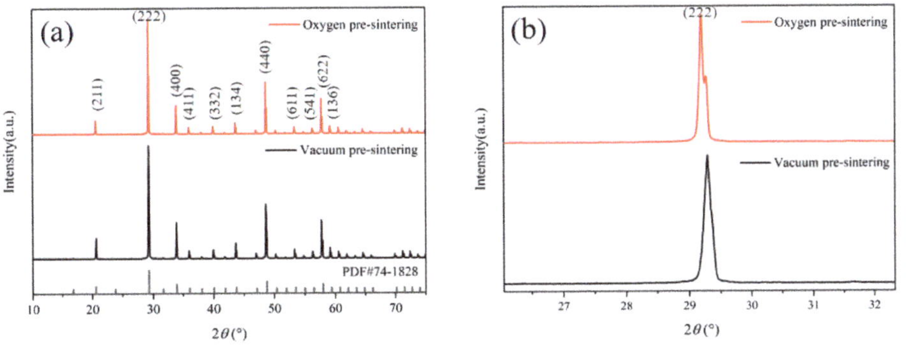

Figure 4. XRD patterns of Y_2O_3 ceramics in different pre-sintering atmospheres in ranges of (**a**) 10°–90° and (**b**) 27°–32°.

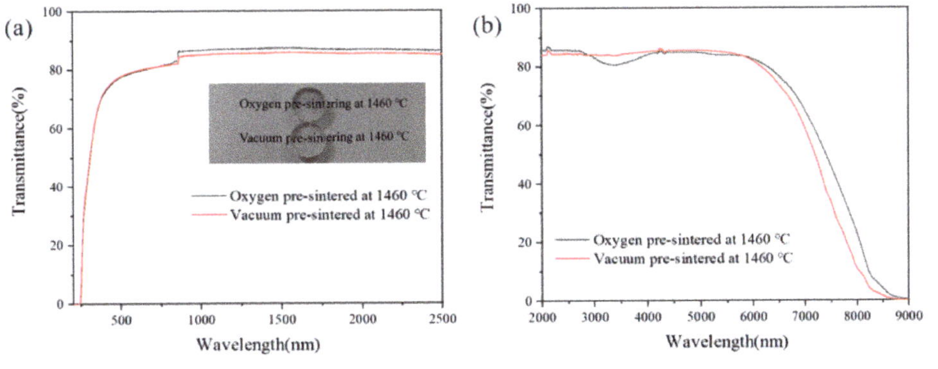

Figure 5. Comparison of the transmittance of Y_2O_3 ceramics in the (**a**) ultraviolet to near-infrared and (**b**) near-infrared to mid-infrared wavelength ranges after vacuum or oxygen pre-sintering followed by HIP.

Table 1. Optical properties of Y_2O_3 transparent ceramics from the literature and this work.

Sample	Sintering Method	Sintering Additive	Transmittance	Ref.
Y_2O_3	VS + HIP	None	83.0%@3–5 μm	[15]
Y_2O_3	AS + HIP	None	81.8%@800 nm	[16]
Y_2O_3	VS	ZrO_2	80%@3–6 μm	[17]
Y_2O_3	OS	ZrO_2	80%@800 nm	[9]
Y_2O_3	VS + HIP	None	81.6%@800 nm 84.0%@3.3 μm	This work
Y_2O_3	OS + HIP	None	81.9%@800 nm 80.8%@3.3 μm	This work

The above reaction is reversible, which effectively explains why hydroxyl groups cannot be removed even by high-temperature calcination under flowing oxygen conditions.

The presence of oxygen molecules increases the concentration of O^{2-}, which, as a product, drives the reaction in the reverse direction. In contrast, under vacuum conditions, the removal of water vapor products is accelerated, which promotes the reaction to proceed in the forward direction.

To investigate the temperature range at which hydroxyl groups begin to decompose during vacuum sintering, Y_2O_3 green bodies were subjected to vacuum heat treatment within the temperature range of 600–1460 °C. To prevent moisture absorption, the samples were immediately densified in a flowing oxygen atmosphere at 1460 °C after the respective vacuum heat treatment, ensuring complete closure of the pores. The samples were then subjected to hot isostatic pressing at 1440 °C to achieve full densification. Figure 6 shows the FTIR spectra of the densified Y_2O_3 samples with different vacuum heat treatment temperatures. It can be observed that after low-temperature (600 °C) vacuum heat treatment, the hydroxyl content in the samples significantly decreased. The above phenomenon indicates that the primary driving force for hydroxyl desorption in Y_2O_3 green bodies is the rapid removal of water vapor in a vacuum environment, rather than the high-temperature decomposition of Y–OH bonds. This finding provides new insights into the desorption behavior of hydroxyl groups in oxide-based infrared optical materials, offering a potential strategy for controlling hydroxyl-related absorption in similar systems.

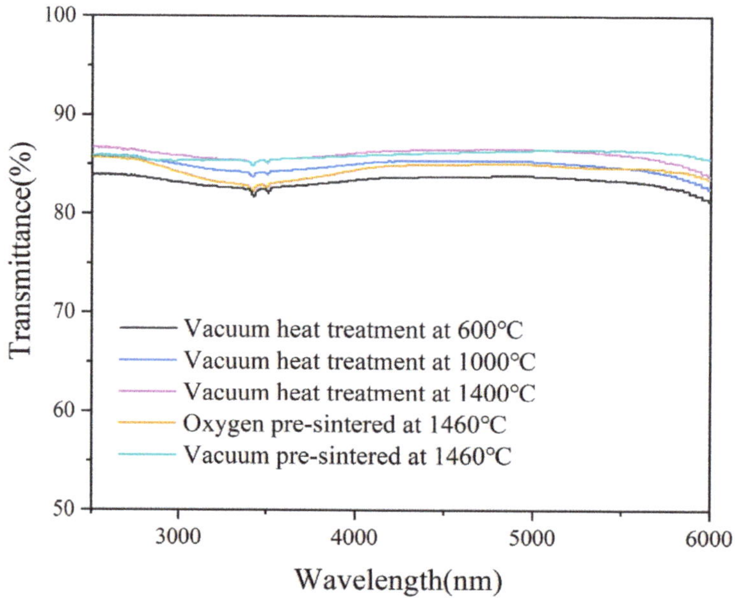

Figure 6. FTIR spectra of the densified Y_2O_3 samples with different vacuum heat treatment temperatures.

Figure 7a illustrates the relationship between the output power and the absorbed pump power for the $Yb:Y_2O_3$ ceramic sample pre-sintered in an oxygen atmosphere and further densified by hot isostatic pressing (HIP) in a simple two-mirror laser cavity. Using a fiber coupled semiconductor laser with a central wavelength of 976 nm, the numerical aperture is 0.22, and the fiber diameter is 105 μm. The pump light output from the laser is collimated through lens F1 (focal length f = 50 mm) and then focused into the ceramic gain medium by lens F2 (focal length f = 125 mm). In order to reduce the thermal effect caused by high-power laser pumping, indium foil was used to tightly wrap the laser ceramic and fix it in the copper heat sink, and the cooling water temperature was set at 20 °C. The resonant cavity structure used in this study is a double mirror flat cavity structure. Among them, the planar input mirror M1 is coated with a two-color dielectric film, which has high transmittance (T > 98%) in the pump light band and high reflectivity (R > 99.9%) in the laser band. The output coupling mirror OC is a planar structure [18]. The $Yb:Y_2O_3$ ceramic sample had dimensions of 3 mm × 3 mm × 3 mm and was used without an anti-reflection coating. The threshold absorbed pump powers were 1.5 W, 3.0 W, and 3.0 W with the output couplers (OCs) of 5%, 10%, and 15%, respectively. With a 10% OC, a maximum output power of 12.85 W was achieved at an absorbed pump power of 38.7 W, corresponding to a slope efficiency of 84.4%. The laser efficiency and output power were comparable to those obtained from vacuum-sintered $Yb:Y_2O_3$ ceramics [18]. This result indicates that the presence of hydroxyl in $Yb:Y_2O_3$ ceramics does not adversely affect laser performance in the 1 μm wavelength range. Therefore, it is reasonable to believe that oxygen-atmosphere sintering might be a promising approach for fabricating rare-earth-doped Y_2O_3 ceramics for laser applications below 2.9 μm wavelengths.

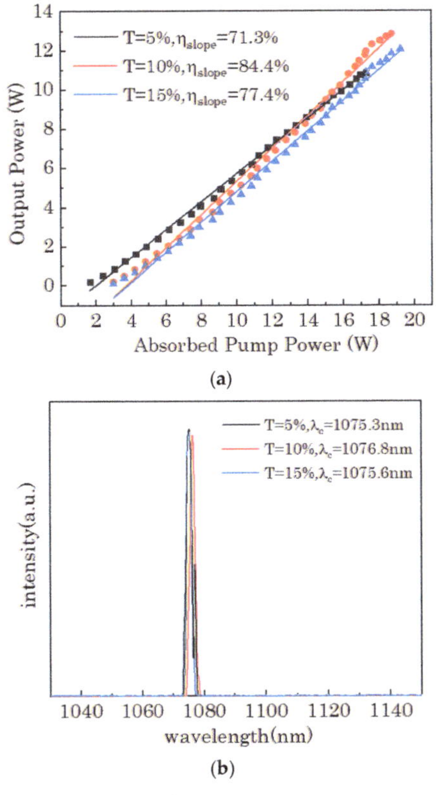

Figure 7. (**a**) Laser performance of the $Yb:Y_2O_3$ ceramic sample pre-sintered in a flowing oxygen atmosphere and subsequently processed by HIP and (**b**) typical laser output spectra for different output couplers.

4. Conclusions

This paper investigated the effects of two sintering atmospheres—vacuum and flowing oxygen—on the sintering behavior of Y_2O_3 ceramics, focusing on densification characteristics and microstructural evolutions. The results indicate that, compared with vacuum sintering, oxygen-atmosphere sintering effectively suppresses grain growth during the final sintering stage and enhances microstructural uniformity. At equivalent sintering densities, oxygen-sintered samples exhibit significantly smaller average grain sizes than those sintered under vacuum. In the 2.8–4.1 μm mid-infrared range, oxygen-sintered samples show distinct hydroxyl absorption peaks. In contrast, vacuum sintering effectively eliminated hydroxyl groups, thereby reducing mid-infrared absorption. Despite the presence of hydroxyl groups, the laser performance of Yb:Y_2O_3 ceramics densified in an oxygen atmosphere demonstrated comparable laser efficiency and output power in the 1 μm wavelength range, underscoring their potential for laser applications outside the 2.8–4.1 μm absorption regions. In summary, oxygen sintering offers a more cost-effective and scalable route for fabricating laser ceramics, while vacuum sintering is preferable for mid-infrared window applications due to its superior hydroxyl removal and reduced mid-infrared absorption.

Author Contributions: Conceptualization, X.W., D.X., Y.W. and D.T.; Formal analysis, X.W., D.X., Y.W., J.W., J.M., P.L., J.Z. and D.T.; Investigation, Y.W. and D.T.; Methodology, X.W., D.X., J.W., J.M., P.L. and J.Z.; Supervision, J.M.; Writing—original draft, X.W., D.X., Y.W. and D.T.; Writing—review and editing, X.W., D.X., Y.W., J.W., J.M., P.L., J.Z. and D.T. All authors have read and agreed to the published version of the manuscript.

Funding: This research was funded by the National Key R&D Program of China (No. 2022YFB3605800); the National Natural Science Foundation of China (Nos. 62105130, 62475108); the Research Capability Elevation Program of Guangdong Province (No. 2022ZDJS116); the National Science Foundation of Top Talent of SZTU (No. GDRC202302).

Institutional Review Board Statement: Not applicable.

Data Availability Statement: The original contributions presented in this study are included in the article. Further inquiries can be directed to the corresponding author(s).

Conflicts of Interest: The authors declare no conflict of interest.

References

1. Liu, Z.; Ikesue, A.; Li, J. Research progress and prospects of rare-earth doped sesquioxide laser ceramics. *J. Eur. Ceram. Soc.* **2021**, *41*, 3895–3910. [CrossRef]
2. Ikesue, A.; Aung, Y.L. Ceramic laser materials. *Nat. Photonics* **2008**, *2*, 721–727. [CrossRef]
3. Ren, C.; Huang, W.; Xie, H.; Wang, F.; Shen, D.; Wang, J.; Zhu, H.; Tang, D. High power and efficient operation of a Ho:Y_2O_3 ceramic laser with over 210 W of output power at 2.1 μm. *Opt. Express* **2022**, *30*, 31407–31414. [CrossRef] [PubMed]
4. Ning, K.; Wang, J.; Luo, D.; Zhang, J.; Dong, Z.L.; Kong, L.B.; Tang, D.Y. New double-sintering aid for fabrication of highly transparent ytterbium-doped yttria ceramics. *J. Eur. Ceram. Soc.* **2016**, *36*, 253–256. [CrossRef]
5. Eilers, H. Fabrication, optical transmittance, and hardness of IR-transparent ceramics made from nanophase yttria. *J. Eur. Ceram. Soc.* **2007**, *27*, 4711–4717. [CrossRef]
6. Pirri, A.; Toci, G.; Patrizi, B.; Vannini, M. An overview on Yb-Doped transparent polycrystalline sesquioxide laser ceramics. *IEEE J. Sel. Top. Quantum Electron.* **2018**, *24*, 1602108. [CrossRef]
7. Ding, M.; Li, X.; Wang, F.; Shen, D.; Wang, J.; Tang, D.; Zhu, H. Power scaling of diode-pumped Er:Y_2O_3 ceramic laser at 2.7 μm. *Appl. Phys. Express* **2022**, *15*, 062004. [CrossRef]
8. Yavetskiy, R.P.; Balabanov, A.E.; Parkhomenko, S.V.; Kryzhanovska, O.S.; Doroshenko, A.G.; Mateychenko, P.V.; Tolmachev, A.V.; Li, J.; Jiang, N.; Gheorghe, L.; et al. Effect of starting materials and sintering temperature on microstructure and optical properties of Y_2O_3:Yb^{3+} 5 at% transparent ceramics. *J. Adv. Ceram.* **2021**, *10*, 49–61. [CrossRef]
9. Huang, Y.; Jiang, D.; Zhang, J.; Lin, Q.; Huang, Z. Sintering of transparent yttria ceramics in oxygen atmosphere. *J. Am. Ceram. Soc.* **2010**, *93*, 2964–2967. [CrossRef]
10. Zhang, L.; Yang, J.; Yu, H.; Pan, W. High performance of La-doped Y_2O_3 transparent ceramics. *J. Adv. Ceram.* **2020**, *9*, 493–502. [CrossRef]
11. Li, Q.; Wang, Y.; Wang, J.; Ma, J.; Ni, M.; Lin, H.; Zhang, J.; Liu, P.; Xu, X.; Tang, D. High transparency Pr:Y_2O_3 ceramics: A promising gain medium for red emission solid-state lasers. *J. Adv. Ceram.* **2022**, *11*, 874–881. [CrossRef]
12. Chen, P.L.; Chen, I.W. Grain boundary mobility in Y_2O_3: Defect mechanism and dopant effects. *J. Am. Ceram. Soc.* **1996**, *79*, 1801–1809. [CrossRef]

13. Seaverson, L.M.; Luo, S.Q.; Chien, P.L.; McClelland, J.F. Carbonate Associated with Hydroxide Sol-Gel Processing of Yttria: An Infrared Spectroscopic Study. *J. Am. Ceram. Soc.* **1986**, *69*, 423–429. [CrossRef]
14. Brindley, G.W.; Nakahira, M. Kinetics of dehydroxylation of kaolinite and halloysite. *J. Am. Ceram. Soc.* **1957**, *40*, 346–350. [CrossRef]
15. Liu, L.; Zhu, Q.; Zhu, Q.; Jiang, B.; Feng, M.; Zhang, L. Fabrication of fine-grained undoped Y_2O_3 transparent ceramic using nitrate pyrogenation synthesized nanopowders. *Ceram. Int.* **2019**, *45*, 5339–5345. [CrossRef]
16. Fu, Z.; Li, X.; Zhang, M.; Zhu, Q.; Li, J.G.; He, J.; Wang, X.A.; Sun, X. Achieving fabrication of highly transparent Y_2O_3 ceramics via air pre-sintering by deionization treatment of suspension. *J. Am. Ceram. Soc.* **2021**, *104*, 2689–2701. [CrossRef]
17. Li, X.; Mao, X.; Feng, M.; Xie, J.; Jiang, B.; Zhang, L. Optical absorption and mechanism of vacuum-sintered ZrO_2-doped Y_2O_3 ceramics. *J. Eur. Ceram. Soc.* **2016**, *36*, 4181–4184. [CrossRef]
18. Li, Q.; Wang, J.; Ma, J.; Ni, M.; Yang, F.; Liu, P.; Lee, K.Y.; Hsiang, H.I.; Shen, D.; Tang, D. Fabrication of high-efficiency Yb:Y_2O_3 laser ceramics without photodarkening. *J. Am. Ceram. Soc.* **2022**, *105*, 3375–3381. [CrossRef]

ceramics

MDPI

Article

Enhancement of Sol–Gel Coatings for Photoprotection of Rosé Wines

Jennifer Moriones [1,2,3,*], Javier Osés [1,4], Pablo Amézqueta [1], José F. Palacio [1,4], Jonathan Fernández De Ara [1] and Eluxka Almandoz [1,2,*]

1 Asociación de la Industria Navarra (AIN), 31191 Cordovilla, Spain; joses@ain.es (J.O.); jfernandez@ain.es (J.F.D.A.)
2 Science Department, Universidad Pública de Navarra (UPNA), 31006 Pamplona, Spain
3 Institute for Advanced Materials and Mathematics (INAMAT2), 31006 Pamplona, Spain
4 Engineering Department, Universidad Pública de Navarra (UPNA), 31006 Pamplona, Spain
* Correspondence: jmoriones@ain.es (J.M.); ealmandoz@ain.es (E.A.)

Abstract: Exposure to short-wavelength light, including UV-A and blue light, can degrade high-value products like rosé wine, which are usually packaged in colourless bottles. This study investigates the optimisation of sol–gel coatings enhanced with UV-absorbing additives (Tinuvin 479 and semaSORB 20109) to provide photoprotection for rosé wines. Coatings with varying additive concentrations (0.5%, 0.75%, 1%, and 1.5%) were applied to glass substrates via spin coating and cured with UV light. Then, optical and mechanical characterisation was performed. The 1.5% concentration semaSORB 20109 bilayer coating demonstrated improved photoprotective properties without compromising colour properties, leading to successful application on glass bottles by spray coating. Accelerated degradation tests confirmed that the optimised coating effectively protected against photodegradation, as indicated by the stability of polyphenol levels and colour parameters in rosé wines. The results suggest that these coatings could be a suitable option for commercial-scale applications, enhancing the light resistance of colourless-bottled products.

Keywords: sol–gel; Tinuvin 479; semaSORB 20109; spin coating; UV-absorbers; photoprotection; rosé wine; anthocyanins

1. Introduction

The commercialisation of rosé wines predominantly involves the use of colourless, transparent bottles. This choice is driven by consumer preferences, which often prioritise the aesthetic appeal of the wine to the protective benefits that coloured bottles offer [1–3]. Coloured bottles can shield the wine from various negative effects induced by photooxidation, a process where light exposure leads to undesirable chemical reactions that deteriorate the wine's quality [4–6]. More precisely, light absorption by anthocyanins at UV (280–400 nm) and blue (360–500 nm) wavelengths induces reactions that alter their chemical structure, affecting their stability and leading to degradation in various forms [7]. Anthocyanins are the most relevant polyphenols contributing to the colour of rosé wine [8], and the mentioned photodegradation reduces their antioxidant properties and changes their colour. Additionally, rosé wines also contain riboflavin (vitamin B2) in their matrix [9]. The exposure of this vitamin to wavelengths from 370 to 450 nm causes its excitation and contributes to the formation of volatile compounds that negatively impact the sensory quality of wines through the development of the so-called light-struck taste [10].

 https://doi.org/10.3390/ceramics8010017

It is widely proven that amber and green glass bottles offer protection against UV-A rays (320–400 nm) and shorter wavelengths within the visible spectrum, whereas clear glass (Flint) bottles do not provide any significant protection in these wavelength ranges [11–13]. In the case of flint bottles, they allow approximately 80% of light with wavelengths above 360 nm to pass through. Green bottles, in contrast, block more light than clear ones, particularly below 520 nm, while amber bottles offer even greater protection, filtering out more light across wavelengths shorter than 520 nm [14]. Despite the superior protective properties of green and amber coloured bottles, their use is often sacrificed to meet marketing tendencies.

Over time, various surface modification techniques have been developed to address such challenges. Among these, the sol–gel process has gained prominence for its ability to create thin, functional coatings on diverse substrates, including ceramics and glass bottles [15,16]. The sol–gel method involves transitioning a colloidal solution (sol) into a solid gel phase, which then forms a uniform and adherent coating [17,18]. This technique allows for the deposition of coatings at relatively low temperatures and offers high-purity materials, making it advantageous for many applications [19,20].

One key advantage of sol–gel coatings is their remarkable versatility. Their adaptable properties enable the incorporation of different additives to introduce specific functionalities [21]. For instance, sol–gel coatings can be engineered to exhibit photocatalytic or UV-absorbing properties, enhancing their ability to protect against light-induced damage [22]. In the context of wine preservation, integrating UV-absorbing additives into sol–gel coatings presents a viable solution for protecting colourless bottles from photooxidation.

This study explores the potential of using sol–gel technology to address the challenge of protecting rosé wines in colourless bottles. In particular, we investigate the incorporation of UV-absorbing additives into a commercially available colourless sol–gel solution (Hardrise CHUI-R218.01, Polyrise SAS, Pessac, France) that has previously demonstrated photoprotective effects. Among these additives, both inorganic and organic compounds can be considered.

Inorganic UV absorbers like Titanium Dioxide (TiO_2), Zinc Oxide (ZnO), Barium Oxide (BaO), Strontium Oxide (SrO), and Aluminium Oxide (Al_2O_3) offer UV protection but also provide a number of other properties that make them poor candidates for the purpose of this project. TiO_2 and ZnO are effective in absorbing UV radiation but often cause opacity [23,24]. BaO and SrO provide moderate UV protection with better transparency but are less effective, and Al_2O_3, while durable and stable, offers limited UV protection [25–27].

SemaSORB and Tinuvin are two well-known organic UV absorbers used to protect materials from damage caused by ultraviolet (UV) radiation [22]. SemaSORB is a brand known for manufacturing UV absorbers [28]. These organic compounds that absorb specific UV wavelengths are used in a variety of products, including plastics and coatings. Their purpose is to extend the durability and preserve the properties of materials. It is worth highlighting semaSORB 20109 [(E)-Ethyl 2-(3-ethoxy-4-hydroxybenzylidene)-3-oxobutanoate] (Figure 1A), which structure features a benzylidene group and a β-ketoester group [29]. The structure of the compound includes conjugated systems with π-electrons that can absorb UV light at specific wavelengths [30]. After absorbing UV radiation, the additives go to an excited state. This absorbed energy is dissipated as heat through non-radiative processes. Tinuvin is another well-known brand that offers products to preserve materials from the damage that produces UV radiation [31]. Specifically, the chemical structure of Tinuvin 479 (6-methylheptyl 2-{4-[4,6-di(4-biphenylyl)-1,3,5-triazin-2-Yl]-3-hydroxyphenoxy}propanoate) (Figure 1B) is based on a benzotriazole ring attached to a phenolic group with a methyl group [32]. This structure is the reason why Tinuvin 479 has specific UV light absorption properties. The core of the molecule is the benzotriazole ring, which is effective in absorbing UV radiation [33]. It also contains nitrogen atoms in its structure that can form π bonds with UV light [34]. The phenolic group attached to the benzotriazole ring contributes to UV light absorption by providing additional bonds for interaction with UV radiation [35]. In the context of sol–gel coatings for wine bottles, both semaSORB 20109 and Tinuvin 479 can be incorporated to impart UV-protective properties. These additives are integrated into the sol–gel solution to develop a coating that blocks UV radiation, reducing the risk of photooxidation of the wine in colourless bottles.

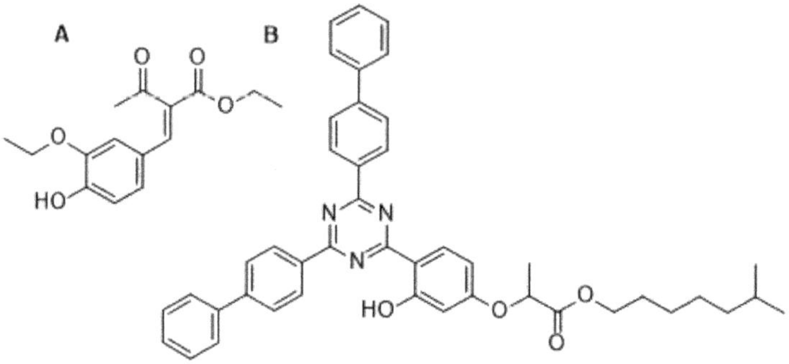

Figure 1. Chemical structure of semaSORB 20109 (**A**) and Tinuvin 479 (**B**).

In contrast to the discussed inorganic absorbers, organic ones like semaSORB 20109 and Tinuvin 479 excel in maintaining the transparency of coatings while delivering targeted UV protection. These organic additives integrate well into sol–gel formulations, providing high-efficiency UV absorption without compromising the visual clarity of the coating [28,32]. They offer greater flexibility in application and performance, making them preferable for situations where aesthetic and functional properties are critical, such as in protecting colourless wine bottles from photooxidation.

Taking into consideration the above information, the primary objective of this project is to enhance a commercial, colourless sol–gel solution with semaSORB 20109 and Tinuvin 479 for wine bottles. The enhanced coating must maintain its colourlessness to align with market preferences while providing improved photoprotection. By achieving this balance, this study aims to offer a practical solution for preserving the quality of rosé wines in transparent bottles, thereby addressing both aesthetic and functional requirements in the wine industry. To verify the effectiveness of the tested coatings, the level of anthocyanins, as well as the colour and the sensory profile of the beverages, will be measured as indicators of protection.

2. Materials and Methods

2.1. Rosé Wine Making and Bottling

Rosé wine was produced using Grenache, Merlot, and Cabernet Sauvignon grape varieties from the 2023 vintage, sourced from OCHOA winery (Olite, Spain). The must from each grape was obtained by separating the free-run juice after a brief maceration period (4–6 h), after which each must was fermented individually. Alcoholic fermentation was carried out at a controlled temperature (16–17 °C) with indigenous yeast. Chips (Nobile Base Chips, NOBILE®, Bordeaux, France) at 1 g/L were used in 30% of the wine during fermentation. Following fermentation, the wines were transferred to separate tanks, where they remained for one month in contact with fine lees. The wines were then blended in a 2:2:1 ratio in a large-capacity tank, clarified with bentonite, and underwent tartaric stabilisation at 4 °C prior to bottling. Minimal sulphur levels were maintained throughout the winemaking process to preserve wine colour. Additionally, nitrogen was introduced to limit oxygen uptake during bottling. The initial sulphite concentration measured 21 ppm. The resulting wine was bottled into coated and uncoated 0.75 L Flint glass bottles for this study.

2.2. Preparation of Sol–Gel Solutions

A commercial sol–gel formula (Hardrise CHUI-R218.01, Polyrise SAS, Canéjan, France) was used as the base. This coating contained a matrix of silica (26%), acrylates (15%), UV absorbers (1.2%), additives (4%), and solvents (53%). Two UV absorbers, Tinuvin 479 (T) (BASF Corporation, Barcelona, Spain) and semaSORB 20109 (S) (se ma Gesellschaft für Innovationen mbH, Coswig, Germany), were added to this base as it is shown in Figure 2. S coatings were doped at 0.5%, 1%, and 1.5%. In the case of T, coatings with 0.5%, 0.75%, 1% and 1.5% were prepared. Higher additive concentrations caused dissolution problems, so they were not studied. The commercial formulation was stored in an opaque container and kept refrigerated until the moment of additive incorporation, after which it was returned to the refrigerator. The formulation with the additive was always used on the same day it was prepared. Each one-litre container of the commercial formulation was used over a one-week period, during which its properties remained unchanged.

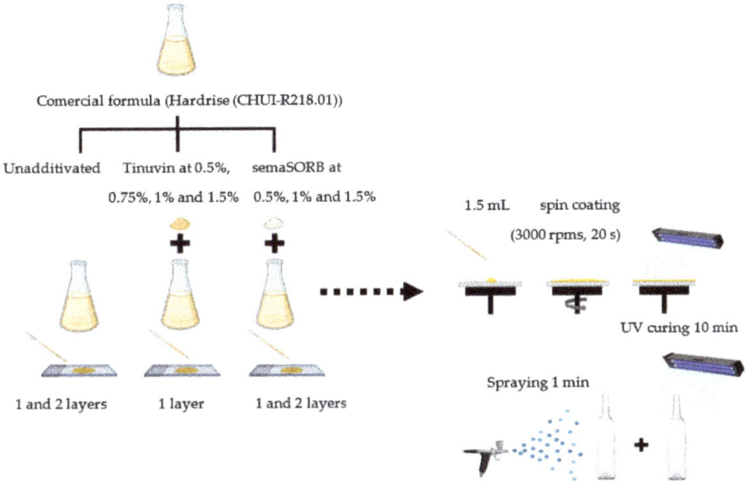

Figure 2. Sample preparation and coating scheme.

2.3. Deposition and Curing of Coatings on Flat Glass (Figure 2)

A total of 1.5 mL of each prepared solution was deposited on flat glass substrates using the spin coating technique (POLOS Advanced new generation ACD300, SPS, Hamburg, Germany) at 3000 rpm of rotation speed for 20 s. S formulations were deposited as single-layer and bilayer coatings. Applying more than two deposition cycles with SemaSORB resulted in poor adhesion of the third layer and an island-like morphology on the glass substrate. That compromised their uniformity and stability, so these coatings were not studied. T coatings were applied as a single layer at the aforementioned concentrations on the same flat substrates as S. In this case, double-layer samples were not studied due to adhesion defects. Additionally, substrates were coated with the base coating as single and double layers to establish a comparison between the different additive concentrations and the base coating.

The samples to be characterised are as follows:

- Samples with no coating (Uncoated)

- Samples coated with the formulation based solely on the commercial Polyrise SAS formulation (control single layer and control bilayer).
- Samples coated with the formulation based on the commercial Polyrise SAS formulation + the UV-filtering additive semaSORB 20109 (0.5% S single layer, 1% S single layer, 1.5% S single layer, 0.5% S bilayer, 1% S bilayer, and 1.5% S bilayer).
- Samples coated with the formulation based on the commercial Polyrise SAS formulation + the UV-filtering additive Tinuvin 479 (0.5% T single layer, 0.75% T single layer, 1% T single layer, and 1.5% T single layer).

*Measurements labelled as "Zero" correspond to the baseline values of different equipment.

The curing of the coatings was performed by exposure to UV light in a UV chamber for 10 min. A high-pressure tungsten filament lamp (ULTRA-VITALUX, Osram, Exeter, NH, USA) was used for this purpose.

2.4. Characterisation of Coatings

2.4.1. Optical Properties

The transmittance properties of sol–gel coatings applied to flat substrates were analysed using a PerkinElmer Lambda 950 UV/VIS/NIR spectrophotometer (PerkinElmer, Inc., Waltham, MA, USA). This instrument also enabled the measurement of CIELab colour coordinates of coatings and the determination of colour differences between coated and uncoated substrates. The spectrophotometric measurements provided quantitative data on the light transmission of coatings across the ultraviolet, visible, and near-infrared regions. By comparing the spectral data of coated and uncoated samples, it was possible to assess the optical impact of the sol–gel layers and quantify changes in colour appearance using the CIELab colour space system [36,37].

2.4.2. Surface Properties

The relevant surface properties for this study include coating thickness, roughness, optical parameters and wettability. Thickness and roughness measurements were conducted using an interferometer (WYKO Rough Surface Tester 500 TM, Veeco Instruments Inc., Plainview, NY, USA). The interferometric analysis allowed for high precision, noncontact measurements of the coating topography, yielding quantitative information on both thickness and surface roughness parameters. The magnified images of the layers were obtained with an interdimensional microscope (LEICA DMI 5000M, LEICA Mikrosysteme Vertrieb GmbH, Wetzlar, Germany) in brightfield. Wettability was analysed using a contact angle measurement device with the sessile drop method (EASY DROP, KRÜSS GmbH, Hamburg, Germany). This instrument is equipped with a high-resolution camera and software (ADVANCE software 1.7.0.8 by KRÜSS) able to detect the shape of the drop once deposited on the surface and calculate the angle, θ. The contact angle measurements offered a means to assess the hydrophobic or hydrophilic nature of the coated surfaces, which is crucial for understanding their interaction with liquids.

2.4.3. Photostability and Durability

To evaluate the long-term effects of light exposure on the designed coatings, a xenon arc chamber (Q-Sun Xe-1-SCE, Q-Lab, Westlake, OH, USA) was employed. This chamber simulates the damage caused by full-spectrum sunlight. The tests were conducted using a window glass filter, producing spectral outputs equivalent to sunlight passing through window glass on the studied flat specimens. The xenon arc chamber provides a controlled environment for accelerated weathering, allowing for the assessment of the photodegradation and colour stability of the coatings under conditions that closely mimic real-world indoor light exposure through windows.

Before and after the test, the colour of the samples was periodically measured with a PerkinElmer Lambda 950 UV/VIS/NIR spectrophotometer (PerkinElmer, Inc., Waltham, MA, USA) in order to know the effect of the tests. At first, these periodic measurements were taken once a day, but over time, the intervals were gradually extended as the aim was to monitor the colour evolution. The process was stopped once a significant colour change was observed. In total, the samples remained in the xenon chamber for 22 days. This colour change was measured by calculating the yellowness of the coatings or Δb from the CIELab scale [37].

2.4.4. Tribological Study

The glass bottles used in wine bottling are coated with a protective layer designed to prevent friction with other bottles during transport, storage or bottling lines. This phenomenon is also known as scuffing and causes visible wear marks or scratches on the glass, often appearing as white or dull grey areas. Measuring the tribological behaviour of the new coatings helps us to verify whether the anti-scuffing properties are maintained in the new coatings. If the coating has a high coefficient of friction, abrasion or surface damage to the bottles may occur. The friction behaviour between coated surfaces was evaluated using a reciprocating tribometer in a ball-on-disc configuration (UMT Tribolab, BRUKER, Billerica, MA, USA). To simulate the interaction between identically coated bottles, both the disc and the counterpart 6 mm glass ball were coated with the same formulation for each test. This experimental setup allowed for the assessment of potential coating damage or failure under conditions to simulate the contact between coated bottles during bottling, handling or transportation. The tribological tests provided quantitative data on the coefficient of friction and wear characteristics, offering insights into the durability and performance of the coatings under simulated contact conditions.

The tribological assessment was conducted using a reciprocating motion test protocol. The experiment consisted of 1000 cycles, with each cycle comprising a 20 mm back-and-forth linear path at a frequency of 1 Hz. A constant normal load of 1 N was applied between the sample and the counterpart throughout the test.

2.5. Scaling up to Bottle

The coating that exhibited the best properties in the characterisation phase was selected for scaling up to bottle application. The coating application on bottles was performed using an airbrush (Suction XTECH 500, SAGOLA, Álava, Spain). Spraying was conducted for 1 min at 2 bars of pressure while the bottle was kept rotating at approximately 0.5 m from the tip of the airbrush. The curing process of each layer was carried out in the UV chamber for 10 min with the bottle in rotation around the lamp.

2.6. Photoprotection Study

To assess the photoprotective capacity of the coatings on wine, an accelerated UV degradation apparatus was utilised [6]. This system comprises 24 high-pressure tungsten filament lamps (ULTRA-VITALUX®, Osram, Munich, Germany) arranged in a 6 × 4 grid. The total irradiance within the chamber is equivalent to natural sunlight at the Earth's surface when the sun is at its zenith. Throughout the experiment, the temperature of the wine bottles was maintained below 26 °C.

For the selection of irradiation times for the experiment, a study was conducted using rosé wines in standard colourless bottles at 0, 20, 38, 53, and 67 h. At 38 h, the wine already exhibited defects during sensory evaluation and a significant decrease in anthocyanins and colour. Therefore, this point, which had previously demonstrated damage in an uncoated wine, was chosen to compare wine in an uncoated bottle (as in the preliminary tests) with wine in coated bottles. Subsequently, the analysis of the rosé wine markers defined for this trial was carried out. Untreated wine was kept in darkness throughout this study as control.

2.7. Determination of Total Anthocyanin Content

The total anthocyanin content was quantified using a validated spectrophotometric pH differential method [38]. Samples were diluted at a 1:4 ratio in two buffer solutions, one at pH 1.0 (0.025 M potassium chloride buffer) and another at pH 4.5 (0.4 M sodium acetate buffer). Absorbance readings were then taken at 520 nm and 720 nm for each pH condition. Rosé wine samples were prediluted 1:4 with methanol. Then, the results were calculated using the following formula:

$$\text{Total anthocyanin content (expressed as malvidin-3-glucoside equivalents)} \\ (\text{mg/L}) = (A \times MW \times DF \times 10^3)/(\varepsilon \times P), \quad (1)$$

where A = (A520 nm − A700 nm) at pH 1.0 − (A520 nm − A700 nm) at pH 4.5; MW = molecular weight of malvidin-3-glucoside; DF = dilution factor (4); ε = extinction coefficient, in $L{\cdot}mol^{-1}{\cdot}cm^{-1}$, for malvidin-3-glucoside (28,000); P = path length in cm (1 cm).

2.8. Evaluation of Wine Chromatic Properties

A spectrophotometric analysis was conducted using a high-performance PerkinElmer Lambda 950 UV/VIS/NIR spectrophotometer (PerkinElmer, Inc., Waltham, MA, USA) to examine absorbance profiles across a wavelength range of 250–800 nm. Distilled water served as the blank, while dark-stored wines were used as reference samples. CIELab colour coordinates (L*, a*, b*) were determined through reflectance measurements. To quantify perceptible colour changes, ΔE values were computed based on the Euclidean distance between points in the CIELab colour space, adhering to internationally recognised standards for colour difference evaluation [36,37].

2.9. Sensory Analysis

A sensory panel of eight expert judges performed visual, olfactory, and gustatory assessments for the wine descriptors across different wine samples. The tasting, conducted at ENONATURA (Villatuerta, Spain), adhered to the guidelines of the International Organization of Vine and Wine (OIV). Samples were evaluated immediately after the bottle opening and served at a temperature of 18 °C.

3. Results

3.1. Optical Properties

The study of the transmittance profile is key to evaluating the filtering of the deposited layers. Figures A1 and A2 show the optical characterisation of the layers developed in this work. For the two additives used, the 1.5% concentration gave better light filtering results in the UV-A range (300–400 nm) than the undoped (or control) coating. These results are shown in Figure 3, where it can be seen that the UVA protection is somewhat higher when using the 1.5% T-doped coating in monolayer form than the 1.5% S-doped coating. Furthermore, it illustrates that above 400 nm, this T-doped layer absorbs some light, and its deposition could generate the appearance of colouration on the sample. This does not occur with the 1.5% S-doped layer. The bilayer conformation could only be implemented in the case of S since, in T, the second layer promoted the appearance of bubbles. The addition of a second layer of 1.5% T on top of the sample would certainly promote a worsening of the light absorption in the visible part of the spectrum and generate samples that would not be valid for this study.

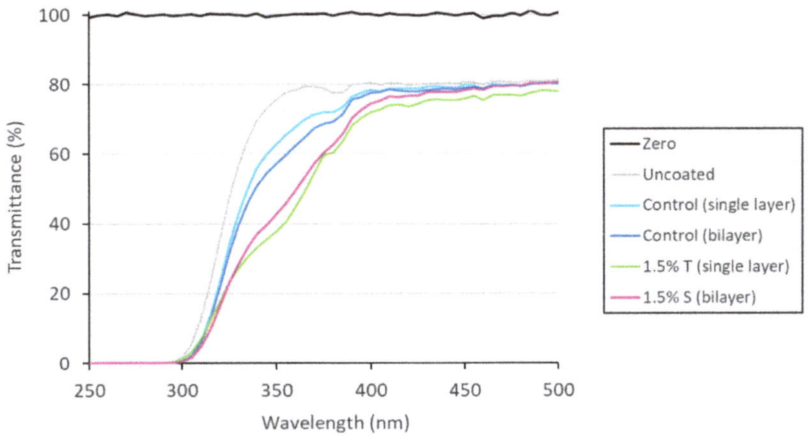

Figure 3. Transmittance profile of the different coatings applied on a flat holder.

Table 1 shows the comparative colour properties of the different additive coatings on flat glass. To begin with, coating the glass sample with the undoped formulation keeps the properties of the substrate unchanged if only one layer is added. By adding the second coating layer, the sample becomes yellowish (increase of b with respect to uncoated). Adding increasing amounts of Tinuvin to this starting formulation causes a gradual increase of the colour component b* and a slight increase of a*. This is accompanied by a slight decrease in the lightness of the sample or L*. In the case of semaSORB doping, the behaviour of Tinuvin is repeated. None of the layers exceeds the critical colour change set to be perceptible to the human eye ($\Delta E \geq 3$) [39], but doped with 1.5% T as a monolayer has a $\Delta E > 2$. At 2, the colour change of a sample is categorised as "visible to most people under normal conditions", and for decorative applications, this can be critical.

Table 1. CIELab L*, a*, and b* coordinates and colour difference of the optimised coatings in contrast to an uncoated sample (first row).

Coating	L*	a*	b*	ΔE (v. Uncoated)
Uncoated	92.62 ± 0.02	0.83 ± 0.01	1.66 ± 0.03	0.00 ± 0.04
Control (1 l.)	92.27 ± 0.03	0.80 ± 0.01	1.83 ± 0.02	0.39 ± 0.01
Control (2 ls.)	92.22 ± 0.04	0.80 ± 0.01	2.51 ± 0.03	0.94 ± 0.02
0.5% T (1 l.)	91.87 ± 0.06	0.91 ± 0.01	3.08 ± 0.01	1.61 ± 0.03
0.75% T (1 l.)	91.86 ± 0.02	0.91 ± 0.04	3.10 ± 0.01	1.63 ± 0.01
1% T (1 l.)	91.42 ± 0.01	0.90 ± 0.03	3.09 ± 0.01	1.87 ± 0.01
1.5% T (1 l.)	91.41 ± 0.05	0.91 ± 0.02	3.34 ± 0.03	2.07 ± 0.08
0.5% S (1 l.)	92.00 ± 0.03	0.85 ± 0.03	2.75 ± 0.02	1.25 ± 0.05
1% S (1 l.)	92.10 ± 0.02	0.93 ± 0.02	3.09 ± 0.01	1.53 ± 0.01
1.5% S (1 l.)	91.94 ± 0.02	0.90 ± 0.03	3.35 ± 0.01	1.82 ± 0.01
0.5% S (2 ls.)	92.17 ± 0.02	0.89 ± 0.01	2.83 ± 0.03	1.26 ± 0.01
1% S (2 ls.)	92.00 ± 0.01	1.00 ± 0.01	3.00 ± 0.03	1.49 ± 0.01
1.5% S (2 ls.)	91.70 ± 0.10	1.01 ± 0.01	3.09 ± 0.05	1.71 ± 0.07

3.2. *Surface Properties (Thickness, Roughness and Wettability)*

The surface properties of the coated samples are of great importance. Roughness impacts the coating's aesthetics, functionality, and durability, influencing tactile and visual perception, optical properties, and resistance to wear. Uniform thickness is crucial to maintain the coating's properties, while inappropriate values may compromise protection or create defects. Finally, controlling wettability enhances the product's appearance and perception, with features such as hydrophobicity being associated with higher quality. The roughness, thickness and contact angle provided by the different coatings applied are shown in Table 2.

Table 2. Surface roughness properties (Ra, Rq, Rt, and Rz), thickness, and contact angle of coatings obtained by doping commercial sol gel coatings with S and T.

Coating	Ra (nm)	Rq (nm)	Rz (μm)	Rt (μm)	Thickness (μm)	Contact Angle (°)
Uncoated	9.5 ± 0.6	12.0 ± 0.7	0.2 ± 0.0	0.2 ± 0.0	-	31.8 ± 3.0
Control (1 l.)	19.3 ± 4.3	31.2 ± 6.3	0.7 ± 0.2	1.3 ± 0.3	0.3 ± 0.0	78.6 ± 3.0
Control (2 ls.)	26.8 ± 7.2	51.7 ± 17.8	0.8 ± 0.2	1.5 ± 0.4	0.6 ± 0.0	82.5 ± 2.4
0.5% T (1 l.)	18.6 ± 2.1	30.4 ± 4.4	0.8 ± 0.1	1.2 ± 0.1	0.5 ± 0.0	82.2 ± 4.2
0.75% T (1 l.)	17.5 ± 1.3	31.4 ± 3.5	0.8 ± 0.2	1.5 ± 0.4	0.4 ± 0.1	82.5 ± 1.0
1% T (1 l.)	123.9 ± 6.2	191.1 ± 12.3	1.6 ± 0.4	2.2 ± 0.5	0.8 ± 0.0	82.6 ± 2.4
1.5% T (1 l.)	120.3 ± 8.0	177.8 ± 14.8	2.0 ± 0.2	3.0 ± 0.5	0.7 ± 0.1	86.6 ± 4.1
0.5% S (1 l.)	20.1 ± 4.4	44.8 ± 13.8	1.2 ± 0.2	1.5 ± 0.4	1.0 ± 0.2	80.7 ± 0.3
1% S (1 l.)	35.9 ± 9.8	76.2 ± 16.8	1.4 ± 0.2	1.2 ± 0.5	1.0 ± 0.1	84.2 ± 2.4
1.5% S (1 l.)	20.1 ± 4.0	37.3 ± 9.7	0.9 ± 0.2	1.7 ± 0.5	0.7 ± 0.2	90.2 ± 2.9
0.5% S (2 ls.)	27.7 ± 8.9	71.0 ± 18.7	1.5 ± 0.2	1.6 ± 0.4	1.0 ± 0.0	82.9 ± 1.2
1% S (2 ls.)	31.1 ± 7.1	72.9 ± 16.4	1.5 ± 0.3	1.3 ± 0.5	1.3 ± 0.3	82.6 ± 1.6
1.5% S (2 ls.)	23.3 ± 7.4	59.5 ± 21.9	1.2 ± 0.3	1.2 ± 0.3	1.4 ± 0.2	88.3 ± 2.3

The commercial formulation increases the roughness of flat substrates, but when S is added at different percentages, this value does not vary significantly, and no defects are observed in the layers when studied by microscopy (Figure 4). This occurs both when applying the formulations as a monolayer and as a bilayer. On the other hand, when 1% and 1.5% T are added in monolayer form, the roughness of the coating does increase significantly. This change in values is clearly seen in Ra and Rt. These effects are not evident to the naked eye, but when the layers are studied under the optical microscope, the appearance of irregularities in these coatings can be seen (Figure 5).

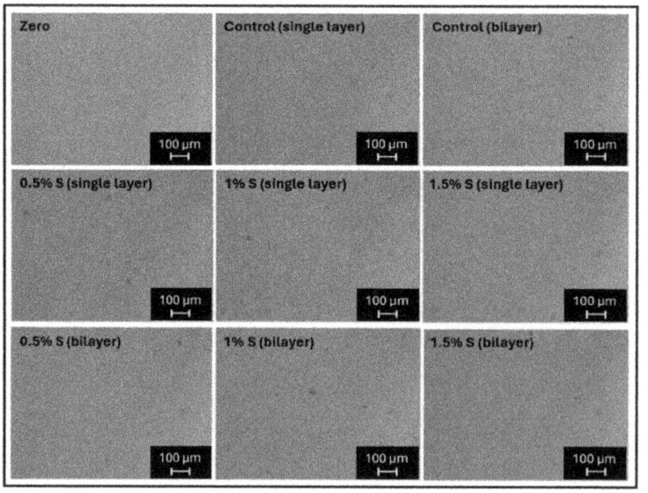

Figure 4. Images obtained by optical microscopy at 10× magnification of coatings with different percentages of S deposited on flat glass sample.

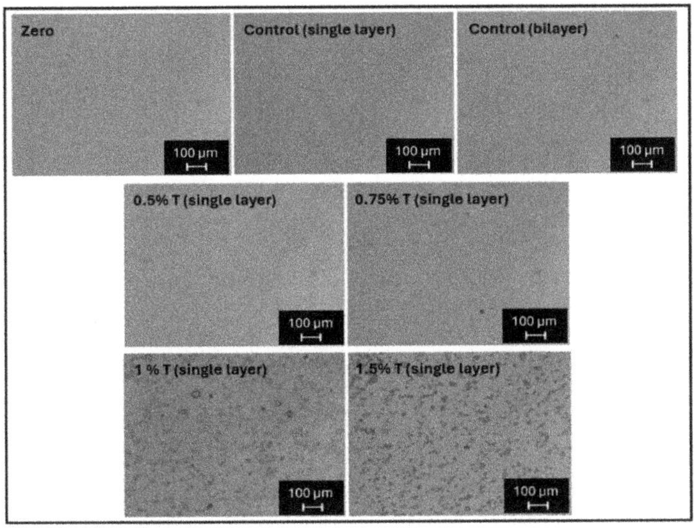

Figure 5. Images obtained by optical microscopy at 10× magnification of coatings with different percentages of T deposited on flat glass sample.

With both absorbents, layers between 0.4 and 1 µm are obtained (Table 2). In the case of T-layers, as the percentage of absorbent increases, the thickness of the layer increases. This can be related to the appearance of increased roughness and flaws mentioned previously at 1% T and 1.5% T. On the other hand, when S is added at different percentages, there is no noticeable increase in layer thickness. The contact angle increases equally in the commercial coating and in the additivated coatings. This is not a significant change, and the additives are not responsible for it.

3.3. Photostability and Durability

The colour variation of the different coatings applied when subjected to degradation tests in Xe chambers is shown in Table 3. In this case, it makes sense to focus this study on the yellowing (Δb) produced by the samples. As the results show, with exposure to these tests on uncoated and undoped samples, the colour of the final sample is not altered towards either yellowing (+Δb) or bluing (−Δb).

Table 3. Yellowing (Δb) of layers with different coatings during exposure to accelerated ageing tests.

Coating	24 h	48 h	72 h	93 h	163 h	333 h	528 h
Uncoated	0.0 ± 0.0	0.0 ± 0.0	0.0 ± 0.0	0.0 ± 0.0	0.0 ± 0.0	0.0 ± 0.0	0.0 ± 0.0
Control (1 l.)	0.0 ± 0.0	0.0 ± 0.0	0.0 ± 0.0	0.0 ± 0.0	0.0 ± 0.0	0.0 ± 0.0	0.0 ± 0.0
Control (2 ls.)	0.0 ± 0.0	0.0 ± 0.0	0.0 ± 0.0	0.0 ± 0.0	0.0 ± 0.0	0.0 ± 0.0	−0.1 ± 0.0
1% T (1 l.)	0.0 ± 0.0	0.0 ± 0.0	0.0 ± 0.0	0.0 ± 0.0	0.1 ± 0.0	0.6 ± 0.2	1.0 ± 0.1
1.5% T (1 l.)	0.1 ± 0.0	0.0 ± 0.0	0.1 ± 0.0	0.1 ± 0.0	0.1 ± 0.0	0.5 ± 0.1	1.1 ± 0.1
1.5% S (1 l.)	−0.1 ± 0.0	−0.3 ± 0.0	−0.4 ± 0.0	−0.4 ± 0.0	−0.4 ± 0.0	−0.4 ± 0.0	−0.4 ± 0.0
0.5% S (2 ls.)	−0.1 ± 0.0	−0.4 ± 0.0	−0.4 ± 0.0	−0.5 ± 0.0	−0.5 ± 0.0	−0.5 ± 0.0	−0.5 ± 0.0
1% S (2 ls.)	−0.4 ± 0.0	−0.4 ± 0.0	−0.4 ± 0.0	−0.4 ± 0.0	−0.4 ± 0.0	−0.5 ± 0.0	−0.6 ± 0.0
1.5% S (2 ls.)	−0.2 ± 0.0	−0.3 ± 0.0	−0.5 ± 0.0	−0.5 ± 0.0	−0.5 ± 0.0	−0.5 ± 0.0	−0.5 ± 0.0

When these tests are carried out on the T-doped samples, no significant colour changes are visible up to 163 h. After that point, the 1% T and 1.5% T coatings become increasingly yellow, giving values of Δb = 1 after 528 h. These Δb values are considered critical for decorative applications such as the objective of this work. In contrast, samples with S-doped coatings start to show signs of blueing in the first 24–48 h. These changes are of a Δb of around −0.5, and as the test time increases, they remain constant. This means that the colour changes of this second family of coatings are imperceptible to slightly noticeable after 528 h.

3.4. Tribological Study

Both 1.5% T single and 1.5% S bilayer coatings were selected for the tribological testing before proceeding to the selection of the coating that would be scaled up to the bottle. Figure 6 shows the results of the coefficient of friction (COF) measurement tests of the layers against glass balls with the same coating. The COF of the uncoated glass sample against an uncoated glass ball is 0.87. When both counterparts are coated with two layers of the base formulation (control), the COF decreases to 0.34. The formulation doped with 1.5% S bilayer shows that the integration of the additive does not significantly alter the COF, as counterparts coated with this formulation yield a value of 0.36. However, when the counterparts are coated with 1.5% T, the resulting COF is 0.40. This test supports the irregularities and roughness variations previously detected in the 1.5% T layer characterisation and is most likely attributed to solubility issues of the additive within the base formulation.

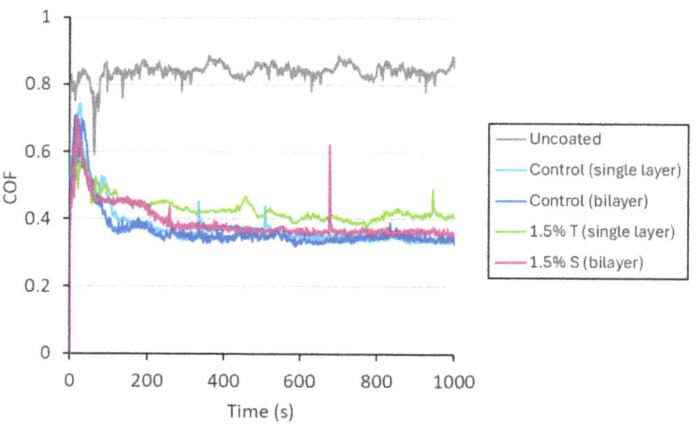

Figure 6. Coefficient of friction of the deposited coatings that offered the most UV light filtering in this work (vs. ball coated with the same coating tested).

As there were no tribological differences, the 1.5% S-doped coating was selected as a bilayer for bottle scale-up since it gave better results after their complete characterisation.

3.5. Results After Bottle Scaling

Figure 7 shows uncoated and coated with two layers of the 1.5% S-doped formulation bottles. As it can be seen in the images, no colouring or defects from the applied coating were visible to the naked eye. When the bottles coated with the S-doped formulation were introduced into the accelerated degradation chamber, problems with the adhesion of the coating to the bottle were detected. The bottles retained the coating, but part of the additive they contained dissolved in the water, turning it yellowish. For this reason, the protection reflected in the oenological parameters may be less than it should be, as part of the additive passed to the water of the irradiation chamber and did not perform its filtering function.

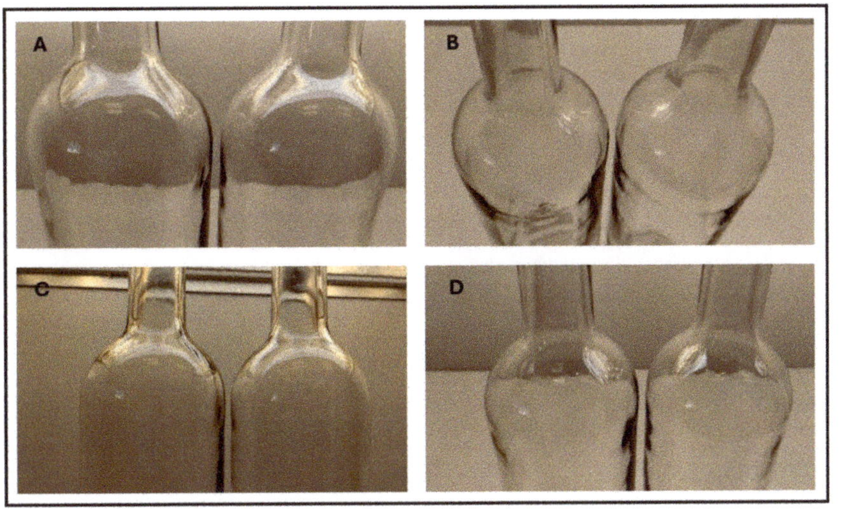

Figure 7. Double coated with the commercial formulation doped with 1.5% S (**A**,**C**) and uncoated bottles (**B**,**D**) from different angles.

3.6. Wine Change with Accelerated Exposure in Colour and Anthocyanins

Rosé wines packaged in coated bottles showed less colour deterioration after 38 h of irradiation than wines in uncoated bottles. This is evident from the increase in the L* coordinate and the reduction in a* and b*, which, as shown in Table 4, were more pronounced in wines stored in uncoated bottles compared to those in coated bottles. Furthermore, the colour change in these wines was greater than 1, which implies that the colour of this wine can be differentiated by a buyer from an unexposed wine. This would not be the case for wines in coated bottles, both undoped and S-doped, as the colour change values are maintained below 1. The small differences in colour between the wines in control-coated bottles and S-doped coated bottles are not significant.

Table 4. CIELab coordinates and colour change of wines exposed to 38 h of accelerated irradiation in different containers.

Index	Unexposed Bottle	Uncoated Bottle-38 h	Control (2 ls.) Bottle-38 h	1.5% S (2 ls.) Bottle-38 h
L*	89.0 ± 0.1	90.1 ± 0.0	89.3 ± 0.1	89.5 ± 0.1
a*	13.6 ± 0.1	12.7 ± 0.0	13.5 ± 0.0	13.2 ± 0.1
b*	13.2 ± 0.1	12.3 ± 0.1	12.5 ± 0.1	12.5 ± 0.1
ΔE	0.0 ± 0.1	1.6 ± 0.1	0.8 ± 0.1	0.9 ± 0.1

On the other hand, the study of the evolution of anthocyanins, shown in Figure 8, gave more information than that collected by the colour indices of the wines. It shows that the irradiation of rosé wines, irrespective of the type of bottle they are in, produces a loss of anthocyanins. The wines in uncoated bottles experienced a loss of 26% of anthocyanins. The wines in coated bottles preserved a greater quantity of these polyphenols, and in this case, it lost 20% of these polyphenols. The study of anthocyanins shows that there are differences between the bottle coated with commercial solution and the one added with semaSORB. The latter preserves more anthocyanins than the other ones and only lost 17% of them.

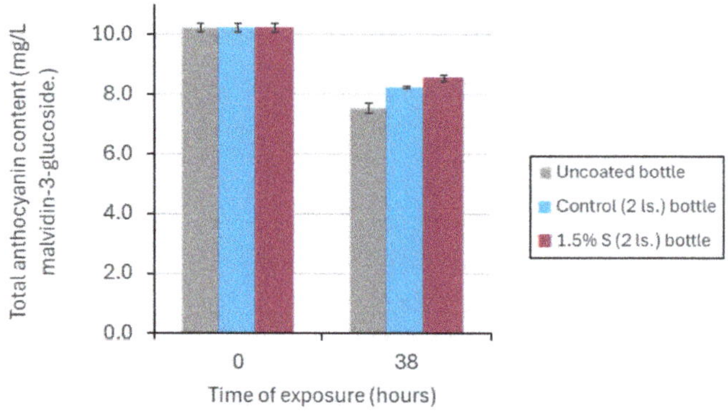

Figure 8. Anthocyanin loss in rosé wines exposed to 38 h of accelerated irradiation in glass containers with different coatings.

3.7. Sensory Analysis

The differences seen in the colour and anthocyanin parameters of the wines translate into the tasting results shown in Table 5. After 38 h of accelerated irradiation, the wines in coated bottles, both S-doped and undoped, present tasting values and descriptors similar to those of the wines not exposed to irradiation. On the other hand, the wines exposed in uncoated bottles have unpleasant and bitter flavours. In addition, visually, a lower layer is visible, which is usually related to a lower number of compounds in the wine matrix.

Table 5. Sensory results with distinctive features appreciated at each stage (visual, olfactory, and taste) in rosé wines packaged in bottles with different coatings after 38-h trials.

Wine Treatment	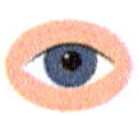			Final Score
Unexposed	More layer (1*)	Flat (1*)	Flat (1*)	74
Uncoated	Less layer (2*)	Flat (3*)	Bitterness (4*), Unpleasant (2*)	69
Control (bilayer)	-	Flat (1*)	-	74
1.5% S (bilayer)	-	Humidity (2*)	-	73

* The number in brackets refers to the number of tasters who detected the mentioned descriptor.

4. Discussion

Protecting the organoleptic and visual properties of rosé wines from light exposure is a key challenge in the wine industry [4–6]. In this study, commercial coatings doped with two additives (T and S) were explored to determine their ability to filter UV light, preserve wine properties, and maintain the aesthetic appearance of the bottles. The results obtained allow the effects of these coatings to be evaluated from multiple perspectives, supporting their potential application at industrial scales.

Variations in additive concentration can influence sol formation by affecting parameters such as hydrolysis and condensation rates, particle size distribution, and overall stability. Higher additive concentrations may lead to increased viscosity and faster gelation, while lower concentrations might result in insufficient cross-linking, impacting the final coating properties.

Comparing the coatings based on the percentage of dopant agent accepted in each case, formulations with 1.5% T were obtained and applied as a single layer due to poor adhesion of a second layer, and formulations with 1.5% S were applied as bilayers. Both dopants are UV absorbers with an absorption peak at 350 nm. The UV filtering performance of T in a single layer is slightly better than that of S in a bilayer. Furthermore, there is no need to add a second layer to achieve the desired effect, making T initially appear as the most suitable option for scaling up.

As the layers were characterised, it became evident that T doping caused a greater colour change compared to S. Under accelerated irradiation, T-doped coatings exhibited yellowing, which, after 528 h of Xe exposure, would become noticeable to customers. This suggests that after storage with exposure to fluorescent light, there is a risk of yellowing becoming detectable to clients.

Although S-doped coatings do not deliver exceptional results in light filtering, when it is applied as a bilayer, it achieves similar effects from the previously mentioned T-doped coatings. Characterisation shows that coatings with this formulation align better with the objective of the project as they exhibit minimal colour changes and result in a more imperceptible final finish. That is the reason why 1.5% S-doped coating as a bilayer was the chosen one for the scale up to bottles.

After 38 h of accelerated irradiation, significant differences in anthocyanin degradation were observed in rosé wines depending on the type of bottle coating used. In uncoated bottles, anthocyanin levels decreased by 26% compared to nonirradiated wine (Figure 7), leading to perceptible changes in colour (Table 4). The commercial coating reduced anthocyanin loss to 20%, which also moderated the changes in the wine's colour coordinates. Finally, the coating doped with 1.5% S provided superior photoprotection, limiting anthocyanin loss to 16%, indicating enhanced efficacy due to the doping.

A relevant study in the literature employed a UV absorber to photoprotect rosé wines [40]. In this work, Tinuvin 1577 was incorporated into low-density polyethylene via hot extrusion, achieving a homogeneous dispersion used to make rosé wine bottles. After 912 h of UV aging in a chamber, rosé wines in uncoated bottles exhibited a 78% anthocyanin loss, whereas bottles with coatings doped with 0.8% Tinuvin 1577 reduced this loss to 71.5%. Although this reduction was significant, the authors concluded that it was insufficient to fully prevent wine degradation. However, when Tinuvin 1577 was combined with Chimassorb 2020, a hindered amine light stabiliser (HALS), anthocyanin loss decreased further to 60%. HALS do not directly absorb UV radiation but are highly effective in neutralising free radicals generated during sunlight exposure, acting synergistically with UV absorbers. This approach appears promising for enhancing the efficacy of the S-doped coatings used in our study.

No other studies have been found that employ UV absorbers specifically to mitigate wine degradation, although research exists for other products, such as milk and oils [41]. De Oliveira [42] reported the incorporation of Tinuvin 234 into PET resins to manufacture oil bottles, achieving a 90% filtration of light at 360 nm. However, after irradiating the oils in bottles with and without the additive, no significant differences were observed in the stability of the oil's compounds. In contrast, in our study, S-doped coatings filtered 37% of the light at 360 nm, which was sufficient to preserve anthocyanins (Figure 7), maintain the wine's colour (Table 4), and improve sensory characteristics during tasting (Table 5). This suggests that the effectiveness of filtration may depend on the critical wavelength responsible for the oxidation of specific compounds in each product.

Other studies, such as that by Coltro et al. [43], demonstrated that using Tinuvin 326 at 0.08% reduced UV light transmission at 360 nm by 90% in doped bottles, providing optimal photoprotection for oils stored under fluorescent lighting for six months. Similarly, De Azeredo et al. [44] reported that Tinuvin 234 improved the oxidative and sensory stability of soybean oil at minimum concentrations of 0.1% during storage at 25 °C. These studies underscore how the concentration and specific type of additive significantly influence the photoprotection provided by coatings.

The loss of semaSORB observed during the introduction of the bottles into the exposure chamber will be addressed by extending the curing time of the sol–gel layers on the bottles. Additionally, based on our results, the layers with 1% S and 1.5% S show very similar properties. Therefore, the percentage of S can be reduced to 1% in the layers to see if this improves the curing of the coatings and prevents the additive from dissolving in the water of the exposure chamber.

The results obtained in this study suggest that the coating doped with S (semaSORB 20109) has strong potential for photoprotecting rosé wines, preserving both their chromatic and sensory attributes. As an optimisation strategy, the incorporation of HALS additives is proposed to enhance photoprotective effects further. Additionally, this strategy could be extended to the coatings doped with 1.5% T developed by our group, which showed a similar 43% light filtration at 360 nm on flat glass (Figure 2).

5. Conclusions

This study demonstrates the effectiveness of sol–gel coatings doped with 1.5% semaSORB 20109 (S) for protecting rosé wines from photodegradation, preserving anthocyanin levels and maintaining sensory properties. Despite the slightly superior UV-filtering performance of Tinuvin 479 (T)-doped coatings, their poor adhesion and noticeable yellowing under accelerated aging tests make them less suitable for commercial applications.

The S-doped bilayer coatings proved to be optically stable, maintaining high transparency and minimal colour, roughness and wettability alterations, which align with market aesthetic preferences. Wines stored in S-doped coated bottles showed reduced anthocyanin loss and better sensory attributes after UV exposure compared to uncoated bottles. Although the preservation of sensory properties cannot be attributed specifically to the use of the S additive, as the bottle coated with the control formulation without additives yields the same results, the anthocyanin measurement does reflect additional protection provided by S.

The scalability of the 1.5% S-doped coating was successfully demonstrated with airbrush application on glass bottles, confirming its potential for industrial implementation. Tests have been carried out to ensure that there will be no issues arising from variations in roughness, friction between bottles, or the durability of the coating during the manufacturing/bottling process. Even so, the curing process and the concentration of additive used still need to be adjusted to achieve the full potential of the coating.

The results highlight that these coatings could be a practical solution for the wine industry to meet consumer demand for transparent bottles while ensuring product quality. Further optimisation, such as incorporating hindered amine light stabilisers (HALS), could enhance photoprotective efficacy. This development represents a promising strategy to extend the shelf life and preserve the sensory characteristics of high-value rosé wines.

6. Limitations

Due to experimental constraints, crystallographic tests are not included in the present study, but we plan to incorporate them in future works.

Author Contributions: Conceptualization, J.M., J.F.D.A. and E.A.; data curation, J.M., P.A. and J.F.P.; formal analysis, J.M., P.A. and J.F.P.; funding acquisition, J.F.D.A. and E.A.; investigation, J.M. and E.A.; methodology, J.O., P.A. and J.F.P.; project administration, E.A.; resources, J.F.D.A. and E.A.; supervision, E.A.; validation, J.M. and E.A.; visualisation, J.M., J.O. and E.A.; writing—original draft, J.M.; writing—reviewing and editing, J.O., J.F.D.A. and E.A. All authors have read and agreed to the published version of the manuscript.

Funding: This research was funded by Proyectos de generación del conocimiento 2021: Ministerio de Ciencia, Innovación y Universidades, grant number PID2021-122675OB-C22 and CNTA Programa EATEX 2023: Fomento de la colaboración en acciones de I+D+i entre comunidades autónomas para centros I+D navarros.

Institutional Review Board Statement: All participants were fully informed that their anonymity was assured, why the research was being conducted, how their data would be used, and if there were any risks involved in participating.

Informed Consent Statement: Verbal informed consent was obtained from the participants. Verbal consent was obtained rather than written to ensure the anonymity of the participants.

Data Availability Statement: The original contributions presented in this study are included in the article. Further inquiries can be directed to the corresponding authors.

Acknowledgments: Authors would like to express their gratitude to Bodegas Ochoa and Polyrise SAS for supplying the rosé wine and the commercial sol–gel formula for this study. Also, for always being available to help us.

Conflicts of Interest: The authors declare no conflicts of interest. Bodegas Ochoa, Polyrise SAS and the funders had no role in the design of this study; in the collection, analysis, or interpretation of data; in the writing of the manuscript; or in the decision to publish the results.

Appendix A

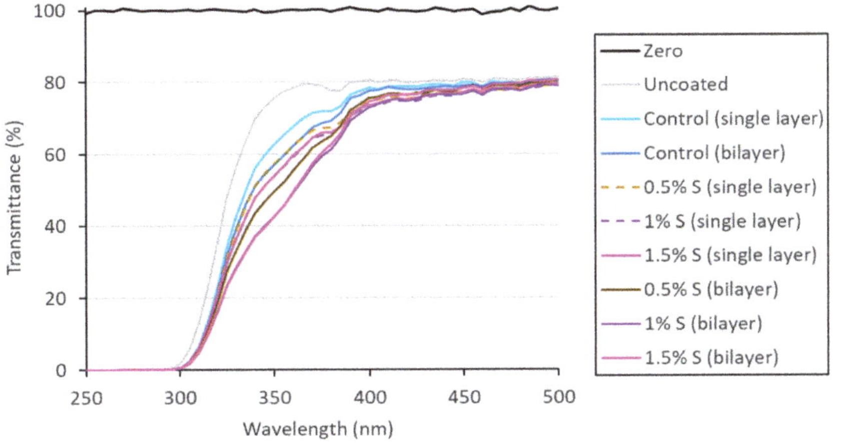

Figure A1. Transmittance profile of the different S percentages coatings applied on a flat holder.

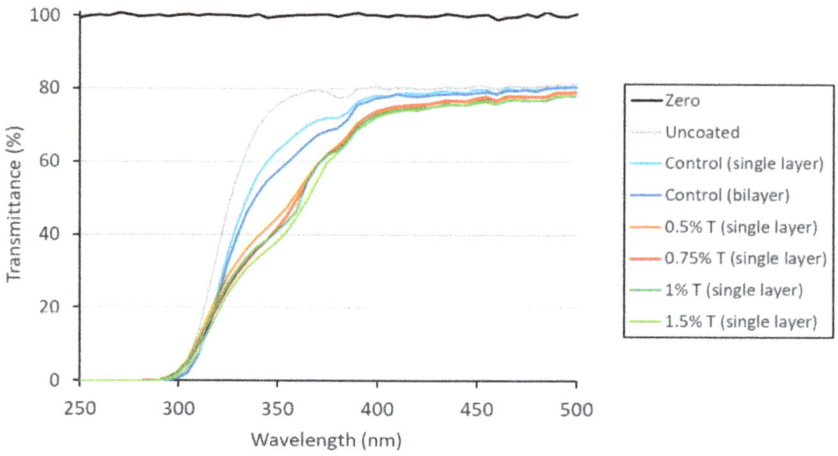

Figure A2. Transmittance profile of the different T percentages coatings applied on a flat holder.

References

1. Ghidossi, R.; Poupot, C.; Thibon, C.; Pons, A.; Darriet, P.; Riquier, L.; De Revel, G.; Mietton-Peuchot, M. The influence of packaging on wine conservation. *Food Control* **2012**, *6*, 302–311. [CrossRef]
2. Maury, C.; Clark, A.C.; Scollary, G.R. Determination of the impact of bottle colour and phenolic concentration on pigment development in white wine stored under external conditions. *Anal. Chim.* **2010**, *660*, 81–86. [CrossRef]
3. Peres, S.; Giraud-Heraud, E.; Masure, A.S.; Tempere, S. Rose Wine Market: Anything but Colour? *Foods* **2020**, *9*, 1850. [CrossRef]
4. Lan, H.; Li, S.; Yang, J.; Li, J.; Yuan, C.; Guo, A. Effects of light exposure on chemical and sensory properties of storing Meili Rosé wine in coloured bottles. *Food Chem.* **2021**, *345*, 128854. [CrossRef]
5. Medina-Plaza, C.; DuBois, A.; Tomasino, E.; Oberholster, A. Effect of storing conditions (lighting, temperature and bottle color) on rosé wine attributes. *Food Chem.* **2024**, *439*, 138032. [CrossRef]
6. Moriones, J.; Jiménez-Moreno, N.; Ancín-Azpilicueta, C.; Fernández de Ara, J.; Navarcorena, B.; Almandoz, E.; Esparza, I. Development of an irradiation equipment to accelerate the degradation of rosé wine in antique green and flint bottles. *Curr. Res. Food Sci.* **2023**, *6*, 100501. [CrossRef]
7. Oliveira, H.; Correia, P.; Pereira, A.R.; Araújo, P.; Mateus, N.; de Freitas, V.; Oliveira, J.; Fernandes, I. Exploring the Applications of the Photoprotective Properties of Anthocyanins in Biological Systems. *Int. J. Mol. Sci.* **2020**, *21*, 7464. [CrossRef]
8. Leborgne, C.; Lambert, M.; Ducasse, M.A.; Meudec, E.; Verbaere, A.; Sommerer, N.; Boulet, J.C.; Masson, G.; Mouret, J.R.; Cheynier, V. Elucidating the Color of Rosé Wines Using Polyphenol-Targeted Metabolomics. *Molecules* **2022**, *27*, 1359. [CrossRef]
9. Mislata, A.M.; Puxeu, M.; Nadal, M.; de Lamo, S.; Mestres, M.; Ferrer-Gallego, R. Influence of different types of LEDs lights on the formation of volatile sulfur compounds in white and rosé wines. *Food Chem.* **2022**, *371*, 131144. [CrossRef]
10. Fracassetti, D.; Tirelli, A.; Limbo, S.; Mastro, M.; Pellegrino, L.; Ragg, E.M. Investigating the Role of Antioxidant Compounds in Riboflavin-Mediated Photo-Oxidation of Methionine: A 1H-NMR Approach. *ACS Omega* **2020**, *5*, 26220–26229. [CrossRef]
11. Dias, D.A.; Clark, A.C.; Smith, T.A.; Ghiggino, K.P.; Scollary, G.R. Wine bottle colour and oxidative spoilage: Whole bottle light exposure experiments under controlled and uncontrolled temperature conditions. *Food Chem.* **2013**, *138*, 2451–2459. [CrossRef]
12. Pajean, G.; Agniel, Y.; Floriot, P. Verre d'emballage et protection UV. *Verre* **2006**, *12*, 23–27.
13. Perscheid, M.; Zuern, F.; Penndorf, G. Effect of light on the quality of beverages. *Fluessiges Obst.* **1979**, *46*, 120–124.
14. Grant-Preece, P.; Barril, C.; Schmidtke, L.M.; Scollary, G.R.; Clark, A.C. Lightinduced changes in bottled white wine and underlying photochemical mechanisms. *Crit. Rev. Food Sci. Nutr.* **2017**, *57*, 4. [CrossRef]
15. Florentino, L.; Conejero, O.; de la Rua, R.B.; Moriones, J.; Fernández de Ara, J.; Fuentes, G.G. Modular UV Curing Sol-Gel Coating for Invisible, Levelling and Easy to Clean Multi-Layer Systems. *Materials* **2023**, *16*, 5442. [CrossRef]
16. Naudin, G.; Ceratti, D.R.; Faustini, M. Sol-Gel Derived Functional Coatings for Optics. In *Sol-Gel Materials for Energy, Environment and Electronic Applications. Advances in Sol-Gel Derived Materials and Technologies*; Pillai, S., Hehir, S., Eds.; Springer: Cham, Switzerland, 2017; pp. 61–99. [CrossRef]
17. Carrera-Figueiras, C.; Pérez-Padilla, Y.; Estrella-Gutiérrez, M.A.; Uc-Cayetano, E.G.; Juárez-Moreno, J.A.; Avila-Ortega, A. Surface Science Engineering through Sol-Gel Process. In *Applied Surface Science*; IntechOpen: London, UK, 2019. [CrossRef]
18. Sopan Mahato, S.; Mahata, D.; Panda, S.; Mahata, S. Sol-Gel Science and Technology in Context of Nanomaterials—Recent Advances. In *Sol-Gel Method-Recent Advances*; IntechOpen: London, UK, 2023. [CrossRef]

19. Jeffrey Brinker, C.; Scherer, G.W. (Eds.) Applications. In *Sol-Gel Science: The Physics and Chemistry of Sol–Gel Processing*; Academic Press: London, UK, 1990; Chapter 14; pp. 838–880. [CrossRef]
20. Hench, L.L.; West, J.K. The sol-gel process. *Chem. Rev.* **1990**, *90*, 33–72. [CrossRef]
21. Sakka, S.; Yoko, T. Sol-gel-derived coating films and applications. In *Chemistry, Spectroscopy and Applications of Sol-Gel Glasses*; Structure and Bonding; Reisfeld, R., JJørgensen, C.K., Eds.; Springer: Berlin, Heidelberg, 1992; Volume 77. [CrossRef]
22. Mahltig, B.; Böttcher, H.; Rauch, K.; Dieckmann, U.; Nitsche, R.; Fritz, T. Optimized UV protecting coatings by combination of organic and inorganic UV absorbers. *Thin Solid Films* **2005**, *485*, 108–114. [CrossRef]
23. Schneider, S.L.; Lim, H.W. A review of inorganic UV filters zinc oxide and titanium dioxide. *Photodermatol. Photoimmunol. Photomed.* **2019**, *35*, 442–446. [CrossRef]
24. Johansson, W.; Peralta, A.; Jonson, B.; Anand, S.; Österlund, L.; Karlsson, S. Transparent TiO_2 and ZnO Thin Films on Glass for UV Protection of PV Modules. *Front. Mater.* **2019**, *6*, 259. [CrossRef]
25. Pacheco, M.H.; Gibin, M.S.; Silva, M.A.; Montagnini, G.; Viscovini, R.G.; Steimacher, A.; Pedrochi, F.; Zanuto, V.S.; Muniz, R.F. BaO–reinforced SiO_2–Na_2O–Ca(O/F_2)–Al_2O_3 glasses for radiation safety: On the physical, optical, structural and radiation shielding properties. *J. Alloys Compd.* **2023**, *960*, 171019. [CrossRef]
26. Kavgacı, M.; Yaykaşlı, H.; Eskalen, H.; Kavaz Perişanoğlu, E.; Kalecik, S.; Yılmaz, R.; Tunç, H. An experimental analysis of the effects of SrO on the mechanical, structural, optical, and nuclear radiation shielding properties of barium borate glasses. *Opt. Mater.* **2024**, *149*, 114975. [CrossRef]
27. Sultana, A.; Kalsoom, U.E.; Abbas Ashraf, G.; Mushtaq, F.; Hammad ul Haq, M.; Mahmood, S.; Ajmal, Z.; Awwad, F.A.; Ismail, E.A.A.; Hassan, N.; et al. Synthesis of novel PdO@Al_2O_3 nanocomposite for energy storage devices, photodegradation, and other applications. *J. Alloys Compd.* **2024**, *1003*, 175719. [CrossRef]
28. Mahltig, B.; Leisegang, T.; Jakubik, M.; Haufe, H. Hybrid sol-gel materials for realization of radiation protective coatings—A review with emphasis on UV protective materials. *J. Sol-Gel Sci. Technol.* **2023**, *107*, 20–23. [CrossRef]
29. SemaSORB® 20109: Product Specification. se ma Gesellschaft für Innovationen mbH. 2018. Available online: https://sema-gmbh.com/wp-content/uploads/6750_20109r3_e-.pdf (accessed on 17 December 2024).
30. Cockell, C.S. Ultraviolet radiation, evolution and the π-electron system. *Biol. J. Linn. Soc.* **1998**, *63*, 449–457. [CrossRef]
31. BASF Corporation. Coatings That Stay Looking Good. Available online: https://www.btc-europe.com/fileadmin/user_upload/Downloads/Brochures-Information/Coatings/Tinuvin_Chimassorb_Lignostab_EN.pdf (accessed on 21 November 2024).
32. Tinuvin® 479. Technical Data Sheet. BASF Corporation. 2019. Available online: https://dispersions-resins-products.basf.us/files/technical-datasheets/Tinuvin_479_October_2019_R4_IC_PP.pdf (accessed on 17 December 2024).
33. Pavanello, A.; Gomez-Mendoza, M.; de la Peña O'Shea, V.A.; Miranda, M.A.; Marin, M.L. Degradation of Benzotriazole UV-stabilizers in the presence of organic photosensitizers and visible light: A time-resolved mechanistic study. *J. Photochem. Photobiol. B Biol.* **2022**, *230*, 112444. [CrossRef]
34. Montesdeoca-Esponda, S.; Vega-Morales, T.; Sosa-Ferrera, Z.; Santana-Rodríguez, J.J. Extraction and determination methodologies for benzotriazole UV stabilizers in personal-care products in environmental and biological samples. *TrAC Trends Anal. Chem.* **2013**, *51*, 23–32. [CrossRef]
35. Nakata, H.; Murata, S.; Filatreau, J. Occurrence and Concentrations of Benzotriazole UV Stabilizers in Marine Organisms and Sediments from the Ariake Sea, Japan. *Environ. Sci. Technol.* **2009**, *43*, 6920–6926. [CrossRef]
36. Mokrzycki, W.; Tatol, M. Colour difference Delta E—A survey. *Mach. Graph. Vis.* **2011**, *20*, 383–411.
37. CIE—Commission Internationale de l'Eclairage Central Bureau. *Technical Report Colourimetry*; CIE—Commission Internationale de l'Eclairage Central Bureau: Vienna, Austria, 2004; pp. 17–18.
38. Lee, J.; Durst, R.W.; Wrolstad, R.E. Determination of total monomeric anthocyanin pigment content of fruit juices, beverages, natural colourants, and wines by the pH differential method: Collaborative study. *J. AOAC Int.* **2005**, *88*, 1269–1278. [CrossRef]
39. Zhang, Q.A.; Shen, Y.; Fan, X.H.; García Martín, J.F. Preliminary study of the effect of ultrasound on physicochemical properties of red wine. *J. Foodserv.* **2016**, *14*, 55–64. [CrossRef]
40. Cristofoli, K.; Brandalise, R.N.; Zeni, M. Photostabilized LDPE Films with UV Absorber and HALS as Protection against the Light for Rosé Sparkling Wine. *J. Food Process. Technol.* **2012**, *166*. [CrossRef]
41. Koontz, J.L. *Packaging Technologies to Control Lipid Oxidation In Oxidative Stability and Shelf Life of Foods Containing Oils and Fats*; AOCS Press: Champaign, IL, USA, 2016; Chapter 13; pp. 479–517. [CrossRef]
42. De Oliveira, M.; Arruda, C.; Ogliari, P.; Meinert, E.; Teixeira, E.; Arellano, D.; Block, J. Efecto de la adición del absorbedor UV (Tinuvin 234) sobre la calidad del aceite de soja en envases de polietileno tereftalato (PET). *Grasas Aceites* **2005**, *56*, 245–253.

43. Coltro, L.; Padula, M.; Saron, E.S.; Borghetti, J.; Buratin, A.E.P. Evaluation of a UV absorber added to PET bottles for edible oil packaging. *Packag. Technol. Sci.* **2003**, *16*, 15–20. [CrossRef]
44. De Azeredo, H.M.C.; Faria, J.A.F.; da Silva, M.A.A.P. The Efficiency of TBHQ, β-carotene, Citric Acid, and Tinuvin 234® on the Sensory Stability of Soybean Oil Packaged in PET Bottles. *J. Food Sci.* **2003**, *68*, 302–306. [CrossRef]

ceramics
MDPI

Article

Optimization of Yb:CaF_2 Transparent Ceramics by Air Pre-Sintering and Hot Isostatic Pressing

Xiang Li [1,2], Chen Hu [1,3], Lihao Guo [1,3], Junlin Wu [1,3], Guido Toci [4], Angela Pirri [5], Barbara Patrizi [4], Matteo Vannini [4], Qiang Liu [2], Dariusz Hreniak [6] and Jiang Li [1,3,*]

1 Transparent Ceramics Research Center, Shanghai Institute of Ceramics, Chinese Academy of Sciences, Shanghai 201899, China; 15839789192@163.com (X.L.); huchen@mail.sic.ac.cn (C.H.); guolihao20@mails.ucas.ac.cn (L.G.); wujunlin20@mails.ucas.ac.cn (J.W.)
2 School of Material Science and Engineering, Jiangsu University, Zhenjiang 212013, China; lq88611338@163.com
3 Center of Materials Science and Optoelectronics Engineering, University of Chinese Academy of Sciences, Beijing 100049, China
4 Istituto Nazionale di Ottica, Consiglio Nazionale delle Ricerche, CNR-INO, Via Madonna del Piano 10C, 50019 Sesto Fiorentino (Fi), Italy; guido.toci@ino.it (G.T.); barbara.patrizi@ino.it (B.P.); matteo.vannini@ino.it (M.V.)
5 Istituto di Fisica Applicata "Carrara", Consiglio Nazionale delle Ricerche, CNR-IFAC, Via Madonna del Piano 10C, 50019 Sesto Fiorentino (Fi), Italy; pirri@ifac.cnr.it
6 Institute of Low Temperature and Structure Research, Polish Academy of Sciences, Okólna 2, 50-422 Wroclaw, Poland; d.hreniak@intibs.pl
* Correspondence: lijiang@mail.sic.ac.cn; Tel.: +86-21-69163689; Fax: +86-21-69906700

Abstract: Yb:CaF_2 transparent ceramics represent a promising laser gain medium for ultra-short lasers due to their characteristics: low phonon energy, relatively high thermal conductivity, negative thermo-optical coefficient, and low refractive index. Compared to single crystals, Yb:CaF_2 ceramics offer superior mechanical properties, lower cost, and it is easier to obtain large-sized samples with proper shape and uniform Yb^{3+} doping at high concentrations. The combination of air pre-sintering and Hot Isostatic Pressing (HIP) emerges as a viable strategy for achieving high optical quality and fine-grained structure of ceramics at lower sintering temperatures. The properties of the powders used in ceramic fabrication critically influence both optical quality and laser performance of Yb:CaF_2 ceramics. In this study, the 5 atomic percentage (at.%) Yb:CaF_2 transparent ceramics were fabricated by air pre-sintering and hot isostatic pressing (HIP) using nano-powders synthesized through the co-precipitation method. The co-precipitated powders were optimized by studying air calcination temperature (from 350 to 550 °C). The influence of calcination temperature on the microstructure and laser performance of Yb:CaF_2 ceramics was studied in detail. The 5 at.% Yb:CaF_2 transparent ceramics air pre-sintered at 625 °C from powders air calcined at 400 °C and HIP post-treated at 600 °C exhibited the highest in-line transmittance of 91.5% at 1200 nm (3.0 mm thickness) and the best laser performance. Specifically, a maximum output power of 0.47 W with a maximum slope efficiency of 9.2% at 1029 nm under quasi-CW (QCW) pumping was measured.

Keywords: Yb:CaF_2 transparent ceramics; air pre-sintering; hot isostatic pressing; microstructure; laser performance

1. Introduction

High peak power and ultrafast laser have broad applications in the fields of laser inertial confinement fusion, high-energy accelerators, femtosecond chemistry, and nuclear physics [1–4]. Nd:Glass, Yb:S-FAP, Ti:Al_2O_3, Yb:YAG, and Yb:CaF_2 are popular materials as gain media in ultrafast lasers [5–10]. Among the materials, Yb:CaF_2 has some advantages, such as broad emission band (72 nm) at 1060 nm, long luminescence lifetime of Yb^{3+}, relatively high thermal conductivity, low non-linear index, and negative thermal-optical coefficient [11,12]. In addition, the phonon energy of Yb:CaF_2 is low (320 cm^{-1}), which is beneficial for reducing the probability of non-radiative transitions and the waste heat generation [13,14]. Thanks to these advantages, Yb:CaF_2 single crystal has been applied in the relevant ultrafast pulse laser devices, and has achieved a peak power of 200 TW [15]. However, the mechanical properties of Yb:CaF_2 single crystal are poor, which could result in scratches and cracks on samples during the machining and cutting processes [16]. In contrast, the mechanical properties of Yb:CaF_2 transparent ceramics are much better than Yb:CaF_2 single crystals [17]. Ceramics have the advantages of high concentration uniform doping, low cost, and highly versatile function design compared with single crystals [18–23]. Therefore, Yb:CaF_2 transparent ceramics are promising laser materials and have attracted considerable attention over the last decades.

The hot forming of single crystals was the first method applied to fabricate Yb:CaF_2 ceramics. Basiev et al. fabricated laser-quality CaF_2-SrF_2-YbF_3 ceramics by hot forming and obtained a continuous laser output through the ceramics for the first time in 2008 [24]. Five years later, Akchurin et al. produced Yb:CaF_2 ceramics through hot forming and investigated the relationship between the Yb^{3+} concentration and thermal conductivity [17]. However, hot forming needs single crystals as starting materials, and the micro-hardness and fracture toughness of ceramics were not remarkably enhanced [25]. A substitute fabrication method of Yb:CaF_2 ceramics includes the synthesis of nano-powders and powder sintering. The primary synthesis approach for Yb:CaF_2 nano-powders is the co-precipitation method, known for its simplicity and cost-effectiveness. It has been widely utilized for various RE:CaF_2 nano-powders (RE = rare earth ions) [26–32], yielding powders with consistent particle size and cubic morphology [28,33–35]. However, there is limited literature concerning the optimization of powder morphology and the further study of sintering behavior.

Commonly, the powder sintering method includes vacuum sintering, hot pressing (HP), and hot isostatic pressing (HIP). Recently, Mei et al. fabricated Yb:CaF_2 ceramics by HP and studied the effects of Yb^{3+} doping concentration, sintering aid YF_3 and NaF on Yb:CaF_2 transparent ceramics [33,36–38]. Among our earlier studies, Yb:CaF_2 transparent ceramics were prepared through vacuum sintering combined with HP post-treatment from co-precipitated powders [39]. The HP temperature has been effectively reduced in this way. However, since HP often involves the use of graphite molds, it may not only lead to carbon contamination of ceramics but also limit the fabrication of large quantities of ceramics [29]. In comparison, HIP technology is a proper method for large-scale production of ceramics, and it is frequently utilized to treat pre-sintered Yb:CaF_2 ceramics with certain densities [40–43]. Air pre-sintering is more readily accessible and cost-effective compared to other sintering techniques. In addition, the organic impurities in ceramics can be removed during the pre-sintering process. During the HIP process, uniform pressure is maintained in all directions of the ceramics and provides a large driving force to discharge the pores inside the pre-sintered ceramics [44–46]. Kitajima et al. fabricated laser-quality Yb:CaF_2 ceramics by air pre-sintering combined with HIP post-treatment and achieved the best continuous laser output available for fluoride transparent ceramics [42]. The influence of air pre-sintering temperature and HIP post-treatment on Yb:CaF_2 ceramics have been studied [40,41]. However, the achieved maximum output power and slope efficiency of the ceramics are not meeting expectations. In addition, the properties of the powders play a crucial role in determining the optical quality and laser performance of Yb:CaF_2 transparent ceramics [47,48]. However, there are few studies investigating powders utilized in air pre-sintering, and the comprehensive impact of air calcination temperature on the properties and microstructure of Yb:CaF_2 ceramics remains largely unexplored.

In this paper, 5 at.% Yb:CaF_2 nano-powders were synthesized using the co-precipitation method. These powders were air calcined at different temperatures. Then, fine-grained Yb:CaF_2 transparent ceramics were fabricated by employing air pre-sintering followed by HIP post-treatment using the air-calcined powders. The research focused on examining how air calcination temperatures influenced the microstructure and optical properties of the Yb:CaF_2 samples. Finally, the laser performance of these samples was investigated to evaluate their suitability for laser applications.

2. Materials and Methods

The 5 at.% Yb:CaF_2 nano-particles were synthesized through the co-precipitation method, utilizing the following raw materials: $Ca(NO_3)_2 \cdot 4H_2O$ (99.90%, Sinopharm Chemical Reagent Co., Ltd., Shanghai, China), $KF \cdot 2H_2O$ (99.90%, Sinopharm Chemical Reagent Co., Ltd., Shanghai, China), and Yb_2O_3 (99.99%, Alfa Aesar, Stoughton, MA, USA). Deionized water served as the solvent to create the $Ca(NO_3)_2$ solution and KF solution, respectively. The $Yb(NO_3)_3$ solution was prepared through a process involving the dissolution of Yb_2O_3 powders in hot nitric acid followed by dilution with deionized water. Subsequently, the KF solution was added gradually to the mixed cationic solution containing $Ca(NO_3)_2$ and $Yb(NO_3)_3$, with a titration rate of 6 mL/min. The synthesis aimed to produce 5 at.% Yb:CaF_2 nano-particles, guided by the following formulation:

$$0.95\,Ca(NO_3)_2 + 0.05\,Yb(NO_3)_3 + 2.05\,KF \rightarrow Yb_{0.05}Ca_{0.95}F_{2.05}\downarrow + 2.05\,KNO_3 \quad (1)$$

The mixed solution underwent stirring for 30 min followed by aging for 12 h at room temperature. Subsequently, the precipitation was washed with deionized water and subjected to centrifugation on multiple occasions to eliminate residual NO^{3-} and K^- ions. The resulting precipitation was then stored at −50 °C for 5 h before freeze-drying in the freeze dryer (Scientz-10N, Ningbo Scientz Biotechnology company, Ningbo, China) under 10 Pa pressure for 12 h and then sieved through a 200-mesh screen. The same method was also described in our previous work [43].

The synthesized nano-powders were calcined in air for 2 h at various temperatures, i.e., 350, 400, 450, 500, 550 °C. Then, the uncalcined and calcined powders were initially dry-pressed in the 20 mm diameter die at 20 MPa. Subsequently, cold isostatic pressing (CIP) was applied at 250 MPa to form green bodies. These green bodies were then air-sintered for 2 h at 625 °C. Then, the pre-sintered samples underwent HIP post-treatment at 600 °C for 3 h under 100 MPa in an Ar atmosphere. The acquired 5 at.% Yb:CaF_2 ceramic samples were double-polished into 3.0 mm for further characterizations. The polished samples were chemically etched in HCl solution for 1 min to observe the microstructure.

The phase identification of the synthesized nano-powders was performed using X-ray diffraction (XRD, Model D/max2200 PC, Rigaku, Tokyo, Japan). The average crystallite size (D_{XRD}) of the 5 at.% Yb:CaF_2 powders can be calculated through Scherrer's formula:

$$D_{XRD} = \frac{0.89\lambda}{\Delta(2\theta) \cdot cos\,\theta} \quad (2)$$

where λ denotes CuKα radiation wavelength (λ = 1.54056 Å), peak position denoted by 2θ, and $\Delta(2\theta)$ is the diffraction peak-corrected FWHM.

The powder morphology and ceramic microstructures were observed through scanning electron microscopy (FESEM, Model SU8220, Hitachi, Tokyo, Japan). The average grain sizes of the ceramics were measured utilizing the linear-intercept method from FESEM images [49]. The in-line transmittance of the samples was assessed using the UV-VIS-NIR spectrophotometer (Model Cray-5000, Varian, CA, USA). Laser emission experiments were performed on 5 at.% Yb:CaF_2 ceramic samples using the longitudinally pumped cavity, utilizing a fiber-coupled laser diode emitting at a wavelength of 930 nm

3. Results and Discussion

Figure 1 illustrates powder X-ray diffraction pattern of the 5 at.% Yb:CaF_2 nano-powders synthesized using the co-precipitation method. The diffraction peaks of the uncalcined powders closely coincide with the standard cubic CaF_2 phase pattern (PDF#35-0816, space group: Fm3m), suggesting the absence of secondary phases. The lattice parameter of the sample is determined to be 5.4688 ± 0.0005 Å, with a corresponding theoretical density of the 5 at.% Yb:CaF_2 samples being 3.4418 g/cm^3. The determined average crystalline size of the powders is approximately 30.0 ± 4.0 nm.

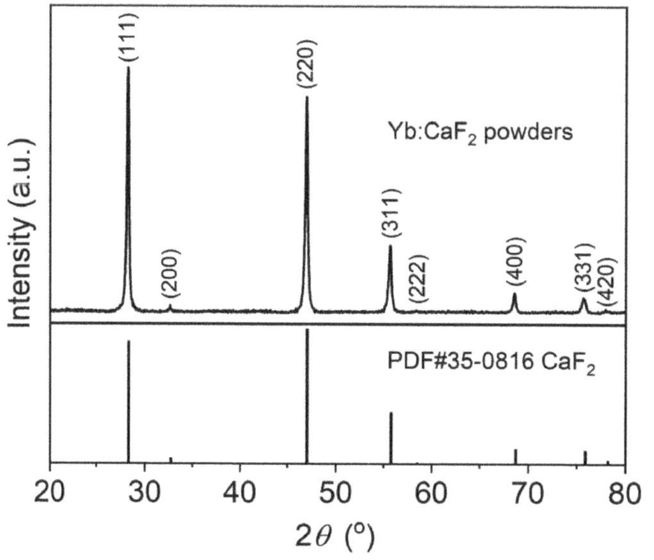

Figure 1. XRD pattern of 5 at.% Yb:CaF_2 nano-particles.

Figure 2 presents the FESEM micrographs illustrating the 5 at.% Yb:CaF_2 nano-powders air-calcined at varying temperatures for 2 h. Prior to air calcination, the powders exhibit a cubic shape and relatively small particle size (see Figure 2a). Upon air calcination at 350 °C, the powder grains undergo further development, transitioning from cubic to spherical in shape. With the increase in calcination temperature to 450 °C, the particle size of the powders enlarges while the degree of agglomeration diminishes, with an average particle size of approximately 60 nm. Upon reaching a calcination temperature of 550 °C, the emergence of sintering necks becomes apparent, and the average particle size of the powders expands to about 200 nm, consequently leading to a reduction in the powder sintering activity (see Figure 2f).

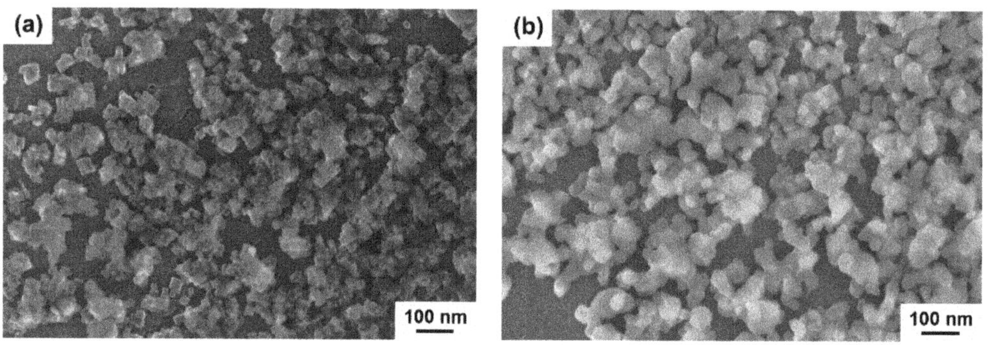

Figure 2. *Cont.*

Figure 2. FESEM micrographs of the 5 at.% Yb:CaF_2 nano-powders air calcined at different temperatures for 2 h: (**a**) without calcination, (**b**) 350 °C, (**c**) 400 °C, (**d**) 450 °C, (**e**) 500 °C, and (**f**) 550 °C.

Figure 3 shows the relative densities and average grain sizes of 5 at.% Yb:CaF_2 ceramics air pre-sintered at 625 °C for 2 h from powders air calcined at different temperatures. The relative density was determined by the Archimedean method. As the calcination temperature rises from 350 °C to 450 °C, the relative densities of pre-sintered ceramics escalate from 96.5% to 98%, contrasting with the 95.3% relative density of ceramics fabricated from uncalcined powders. This elevation in relative density may be attributed to the decreased sintering activity of calcined powders, consequently limiting the formation of pores. When the calcination temperature reaches 550 °C, the relative density of ceramics drops to 90.0%. Meanwhile, it is noticeable that the calcination temperatures of powders exhibit little influence on the average grain sizes of pre-sintered ceramics, which ranges from 250 nm to 300 nm with increasing calcination temperature.

Figure 4 shows the microstructure evolution of pre-sintered Yb:CaF_2 ceramics fabricated from various air calcination temperatures powders. Strip-shaped pores are evident on the surfaces of pre-sintered ceramics derived from uncalcined powders. Following air calcination at 350 °C and 400 °C, the pores on the surfaces of ceramics are mostly isolated pores, characterized by smaller sizes. Compared to strip-shaped pores, smaller isolated pores are more conducive to full discharge during subsequent HIP processes [41]. Upon elevating the calcination temperature to 550 °C, both the size and number of pores on the ceramic surfaces increase, which is consistent with the change of the relative densities of the ceramics (see Figure 3).

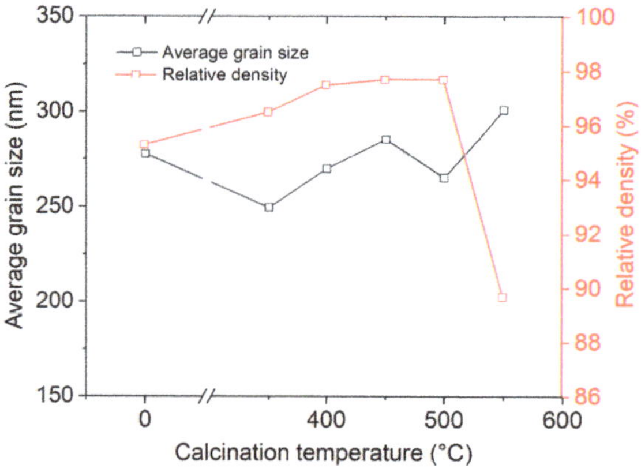

Figure 3. Relative densities and average grain sizes of 5 at.% Yb:CaF_2 ceramics air pre-sintered at 625 °C for 2 h from powders air calcined at different temperatures.

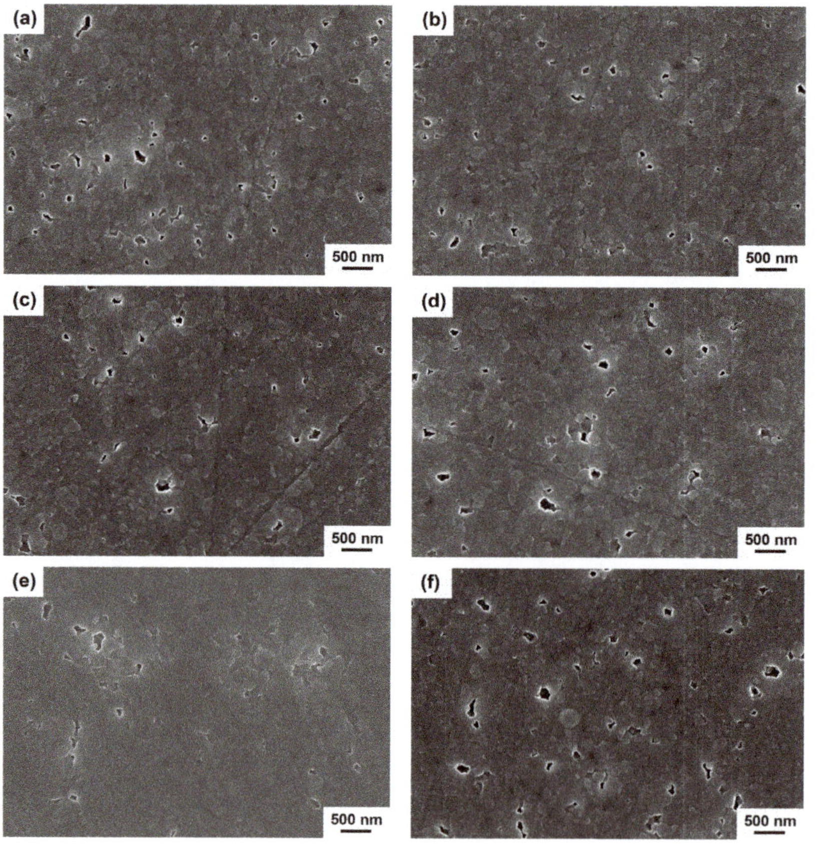

Figure 4. FESEM micrographs of the acid etched surfaces of 5 at.% Yb:CaF_2 ceramics air pre-sintered at 625 °C for 2 h from powders air calcined at different temperatures for 2 h: (**a**) without calcination, (**b**) 350 °C, (**c**) 400 °C, (**d**) 450 °C, (**e**) 500 °C, and (**f**) 550 °C.

Figure 5 presents photographs of 5 at.% $Yb:CaF_2$ transparent ceramics, air pre-sintered at 625 °C, and subsequently HIP post-treated at 600 °C, derived from powders calcined at various temperatures. Notably, ceramic samples fabricated from powders air-calcined at 350~500 °C exhibit remarkable transparency, allowing clear observation of printed text through the samples. This observation suggests that the isolated pores within the pre-sintered ceramics have been effectively eliminated during the HIP treatment. However, the 5 at.% $Yb:CaF_2$ ceramics prepared by 550 °C air calcination powders cannot achieve transparency. This is because the relative density of the corresponding pre-sintering ceramic sample is relatively low, and there are too many internal pores, larger in size, and mostly connected pores (see Figure 4f). During the HIP process, connected pores are difficult to fully discharge and compress compared to isolated pores, forming a large number of scattering centers and ultimately leading to the opacity of the ceramics (see Figure 5). Furthermore, a discernible improvement in transparency is evident in ceramics fabricated from powders air-calcined at 350~500 °C compared to those from uncalcined powders.

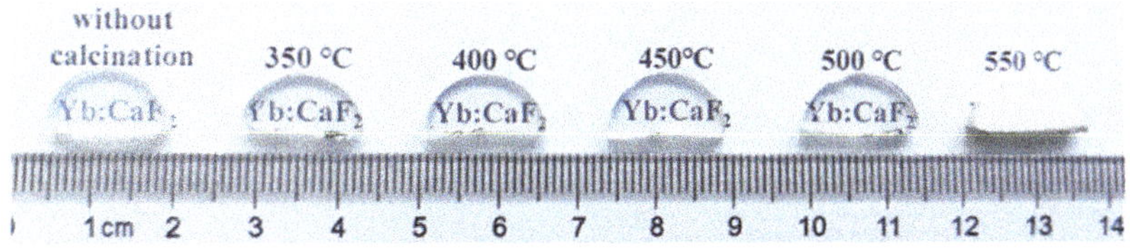

Figure 5. Photograph of the 5 at.% $Yb:CaF_2$ ceramics air pre-sintered at 625 °C and HIP post-treated at 600 °C from powders air calcined at different temperatures, double-face polished to thickness of 3.0 mm.

Figure 6 shows the in-line transmittance curves of 5 at.% $Yb:CaF_2$ transparent ceramics air pre-sintered at 625 °C and HIP post-treated at 600 °C from powders air calcined at different temperatures. Following double-face polishing, the thickness of the 5 at.% $Yb:CaF_2$ transparent ceramics measures 3.0 mm. It is obvious that the ceramic samples prepared by low-temperature air calcination powders demonstrate a significant improvement in the linear transmittance at short wavelengths. Ceramic samples derived from powders calcined at 400 °C exhibit the highest in-line transmittance at 1200 nm, achieving 91.5%, with the corresponding transmittance of 63.5% at 400 nm. Conversely, samples corresponding to powders calcined at 350 °C demonstrate superior in-line transmittance at 400 nm, reaching 65.5%, while registering the transmittance of 91.0% at 1200 nm. Notably, there is a rapid decrease in transmittance within the visible range, likely attributable to residual nano-scale pores within the ceramics (see Figure 7). In accordance with the Rayleigh scattering theory [50,51], light scattering intensity rises as the wavelength decreases. It is likely that pores act as the primary contributors to scattering losses in this material.

Figure 7 shows the FESEM micrographs of acid etched surfaces of 5 at.% $Yb:CaF_2$ transparent ceramics air pre-sintered at 625 °C and HIP post-treated at 600 °C from powders air calcined at different temperatures. It can be seen that a small amount of nano-scale pores are still observed on the surfaces of $Yb:CaF_2$ ceramic samples corresponding to the powders calcined at 350–500 °C, predominantly clustered along the trigeminal grain boundaries. These pores contribute to a decline in the ceramics' linear transmittance, particularly at shorter wavelengths. With escalating calcination temperatures, the grain sizes of ceramics after HIP post-treatment are relatively close. At the same time, the ceramic samples prepared by air calcination of powders at 550 °C exhibit a large number of pores on the surfaces after HIP post-treatment, measuring several hundred nanometers, which accounts for the inability of ceramics to attain transparency after HIP post-treatment (see Figure 7f). The above results indicate that the air calcination of powders at appropriate temperatures can moderately reduce the sintering activity of $Yb:CaF_2$ nano-particles and thus avoid the formation of intracrystalline pores due to rapid grain growth during the pre-sintering process. Ultimately, the optical quality of ceramics after HIP post-treatment can be further improved.

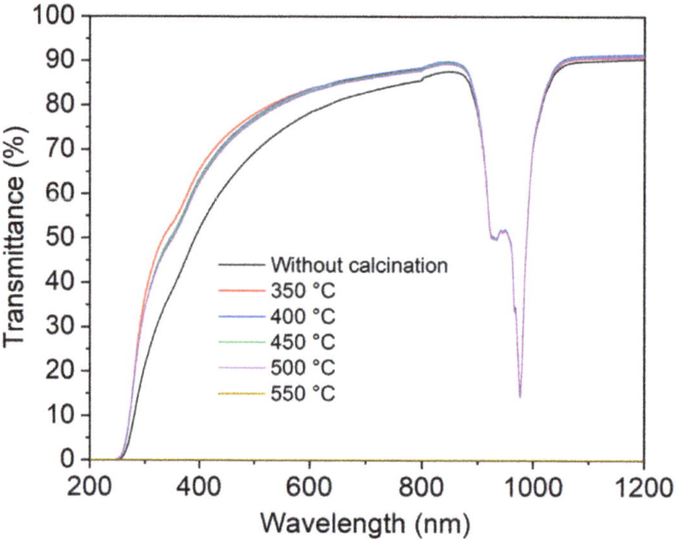

Figure 6. In-line transmittance of 5 at.% Yb:CaF_2 ceramics air pre-sintered at 625 °C and HIP post-treated at 600 °C from powders air calcined at different temperatures.

Figure 8 depicts the average grain sizes before and after HIP post-treatment of the samples derived from powders air-calcined at various temperatures. Following HIP post-treatment, the average grain size of all ceramics measures below 500 nm, with an approximate 200 nm increase observed. This phenomenon may arise due to the close proximity between the post-treatment temperature of HIP and the sintering temperature during hot pressing, potentially limiting the sintering driving force necessary for secondary grain growth. Moreover, the findings suggest that combining air pre-sintering with HIP post-treatment is a promising approach for producing high optical quality and fine-grained Yb:CaF_2 transparent ceramics.

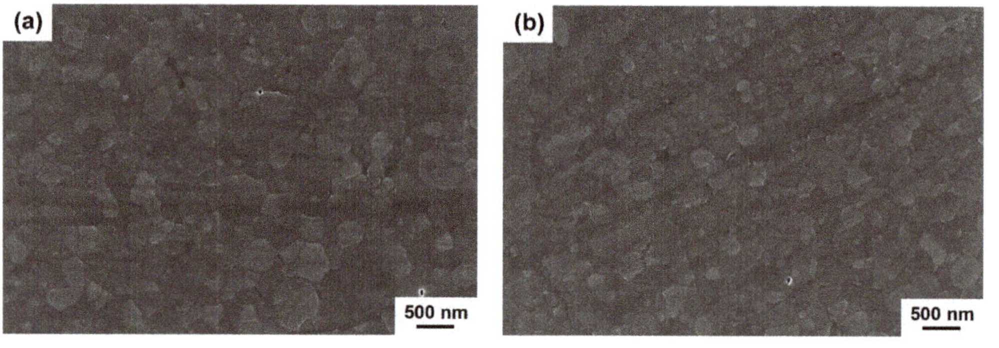

Figure 7. *Cont.*

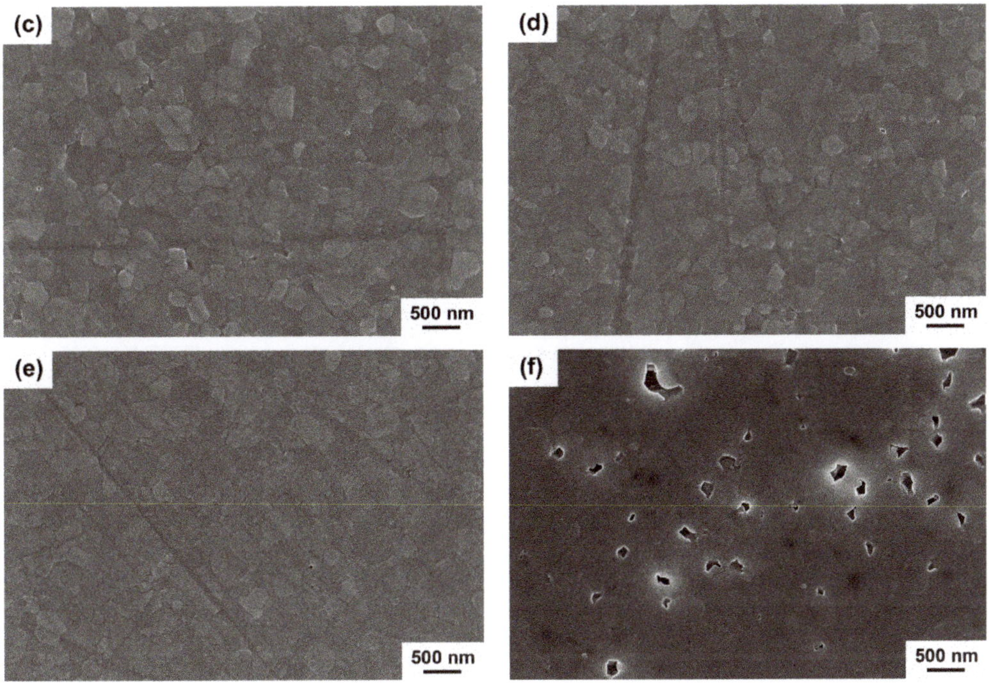

Figure 7. FESEM micrographs of acid etched surfaces of the 5 at.% Yb:CaF_2 ceramics pre-sintered at 625 °C and HIP post-treated at 600 °C from powders air calcined at different temperatures: (**a**) without calcination, (**b**) 350 °C, (**c**) 400 °C, (**d**) 450 °C, (**e**) 500 °C, and (**f**) 550 °C.

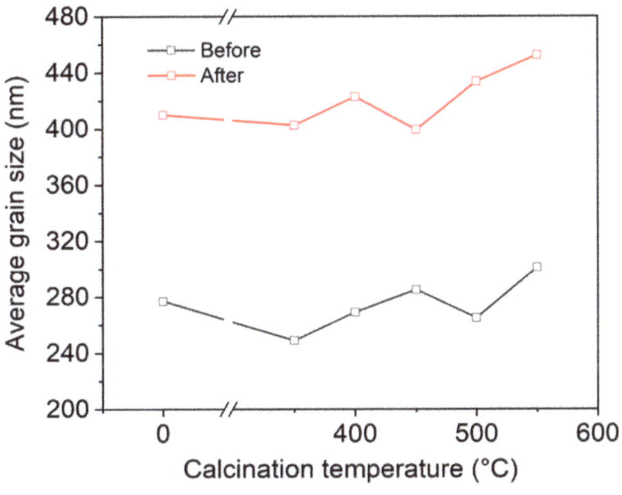

Figure 8. Average grain sizes before and after HIP post-treatment of Yb:CaF_2 ceramic samples pre-sintered at 625 °C and HIP post-treated at 600 °C from powders calcined at different temperatures.

Finally, to conclusively confirm the applicability of the resulting material as the laser medium, we investigated the laser emission properties of the ceramic sample fabricated using the 400 °C air calcined powders, which has a better overall in-line transmittance.

The test setup used to investigate laser performance is shown in Figure 9a. The laser cavity utilizes a V-shaped design with a folding angle of approximately 10°. The cavity configuration includes arms of 55 mm between the high-reflectivity end-mirror (EM) and the folding mirror (FM) and 160 mm between the folding mirror (FM) and the output coupler (OC). This setup ensures the cavity remains stable for optimal operation. Various output coupler mirrors (OCs) are employed, offering transmission rates ranging from T_{oc} = 1.8% to T_{oc} = 12.3%. Heat dissipation is managed by soldering the 2 mm sample with indium onto a copper heat sink, water-cooled to 20 °C. The fiber-coupled laser diode emitting light at λ = 930 nm was used to achieve efficient pumping. The optical fiber core diameter in the system is 105 µm with a numerical aperture of 0.22. This quasi-CW laser operates at a frequency of 10 Hz with a pulse length of 20 ms.

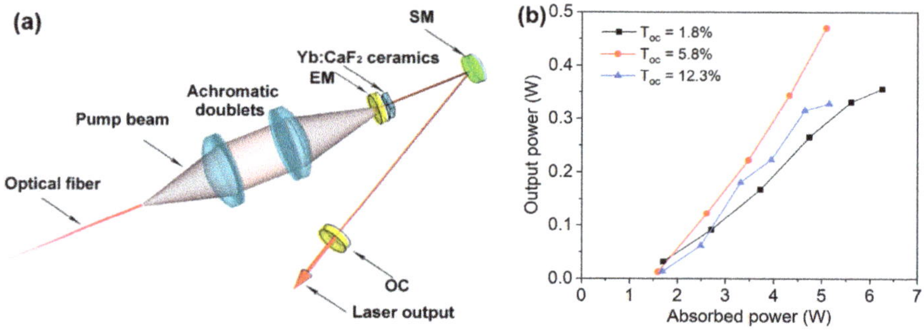

Figure 9. (**a**) Schematic diagram of the laser cavity: EM: End Mirror; SM: Spherical Mirror; OC: Output Coupler Mirrors and (**b**) Laser output power vs. absorbed pump power for different values of the output coupler mirror transmission.

The laser performance of the hot-pressed ceramic sample fabricated from powders air calcined at 400 °C and HIP post-treated at 600 °C is as shown in Figure 9b. Under the QCW pumping condition, the maximum output power is 0.47 W achieved with T_{OC} = 5.8%, which corresponds to the optical-to-optical efficiency (η_o) of 8.1%, and the slope efficiency (η_s) reaches its peak of 9.2%. Detailed parameters of the laser emission can be found in Table 1.

Table 1. Main laser emission parameters of ceramic sample fabricated using the 400 °C air calcined powders with slope efficiency being calculated with respect to the absorbed pump power.

Output Coupler Transmission/%	Maximum Power/W	Slope Efficiency/%	Optical Efficiency/%	Lasing Wavelength/nm
1.8	0.36	7.6	5.7	1028.9
5.8	0.47	9.2	8.1	1029.0
12.3	0.33	8.9	6.4	1030.0

4. Conclusions

Using the co-precipitation method, 5 at.% $Yb:CaF_2$ nano-powders were successfully synthesized. The co-precipitated powders were then air calcined at different temperatures. When the calcination temperature increases from 350 to 550 °C, powder average particle size increases from 40 nm to 200 nm, transitioning from cubic to spherical in shape. The air pre-sintered ceramics corresponding to the powders calcined at 400 °C exhibit the highest relative density of 98.0%. Through HIP post-treatment (600 °C × 3 h, 100 MPa Ar), the pores in the samples are effectively eliminated, and the average grain sizes of the samples through HIP post-treatment show a slight increase, ultimately reaching about 450 nm. Moreover, the 5 at.% $Yb:CaF_2$ transparent ceramics pre-sintered at 625 °C and HIP post-treated at 600 °C corresponding to the powders calcined at 400 °C exhibit good optical quality, reaching the in-line transmittance of 91.5% at 1200 nm. Finally, the laser performance of the ceramic sample produced using the 400 °C air calcined powders was tested under QCW pumping condition. The maximum output power of 1.51 W was measured by closing the laser cavity with T_{oc} = 5.8%. The highest slope efficiency was 9.2% with T_{oc} = 18.1%, which corresponds to an optical-to-optical efficiency η_o = 8.1%.

Author Contributions: Conceptualization, J.L.; validation, C.H., J.W. and G.T.; formal analysis, X.L., L.G., A.P., B.P., M.V. and D.H.; investigation, X.L., L.G., G.T. and J.L.; resources, J.L.; writing—original draft, X.L. and L.G.; writing—review & editing, C.H., J.W., G.T., A.P., B.P., M.V., Q.L., D.H. and J.L.; supervision, Q.L. and J.L. All authors have read and agreed to the published version of the manuscript.

Funding: This work was supported by the National Key R&D Program of China (Grant No. 2021YFE0104800), the International Partnership Program of Chinese Academy of Sciences (Grant No. 121631KYSB20200039), National Center for Research and Development (Contract No.WPC2/1/SCAPOL/2021), and the Chinese Academy of Sciences President's International Fellowship Initiative (Grant No. 2024VEA0014).

Institutional Review Board Statement: Not applicable.

Informed Consent Statement: Not applicable.

Data Availability Statement: The original contributions presented in the study are included in the article, further inquiries can be directed to the corresponding author.

Acknowledgments: The authors would like to thank Lexiang Wu (Transparent Ceramics Research Center, Shanghai Institute of Ceramics, Chinese Academy of Sciences, Shanghai 201899, China) for operating the CIP equipment.

Conflicts of Interest: The authors declare no conflicts of interest.

References

1. Guo, F.F.; Rong, W.L.; Chen, L.J.; Hu, F.A.; Wang, S.P.; Tao, X.T.; Gao, Z.L. Novel laser crystal Nd:LiY(MoO_4)$_2$: Crystal growth, characterization, and orthogonally polarized dual-wavelength laser. *Cryst. Growth Des.* **2024**, *24*, 1421–1428. [CrossRef]
2. Ji, Y.Y.; Hu, M.; Xv, M.M.; Li, H.Z.; Gao, L.; Li, Q.; Bi, M.H.; Zhou, X.F.; Pan, S.Q.; Liu, C. Exploring the spatial hole burning effect on the mode-locking characteristics of self-mode-locked Nd:YVO_4 lasers. *Opt. Commun.* **2023**, *549*, 129883. [CrossRef]
3. Zhang, L.X.; Hu, D.J.; Snetkov, I.L.; Balabanov, S.; Palashov, O.; Li, J. A review on magneto-optical ceramics for Faraday isolators. *J. Adv. Ceram.* **2023**, *12*, 873–915. [CrossRef]
4. Dong, J.S.; Wang, Q.G.; Xu, J.; Xue, Y.Y.; Wang, W.D.; Cao, X.; Tang, H.L.; Wu, F.; Luo, P. Growth and spectral properties of Ho,Y:CaF_2 crystal grown with porous crucible TGT method. *J. Synth. Cryst.* **2022**, *51*, 200–207.
5. Danson, C.N.; Haefner, C.; Bromage, J.; Butcher, T.; Chanteloup, J.-C.F.; Chowdhury, E.A.; Galvanauskas, A.; Gizzi, L.A.; Hein, J.; Hillier, D.I.; et al. Petawatt and exawatt class lasers worldwide. *High Power Laser Sci. Eng.* **2019**, *7*, e54. [CrossRef]
6. Papadopoulos, D.N.; Friebel, F.; Pellegrina, A.; Hanna, M.; Camy, P.; Doualan, J.L.; Moncorgé, R.; Georges, P.; Druon, F.P.H.J. High repetition rate Yb:CaF_2 multipass amplifiers operating in the 100-mJ range. *IEEE J. Sel. Top. Quantum Electron.* **2015**, *21*, 464–474. [CrossRef]
7. Xuan, L.L.; Pisch, A.; Duffar, T. Thermodynamic calculations of Ti ion concentrations and segregation coefficients during Ti:Sapphire crystal growth. *Cryst. Growth Des.* **2022**, *22*, 2407–2416. [CrossRef]
8. Ma, B.Y.; Zhang, W.; Luo, H.; Yuan, F.; Cheng, B.T.; Bai, L.Y.; Tang, Y.; Song, H.Z. Growth of Cr,Yb:YAG single crystals for self-Q-switched monolithic solid-state lasers. *Opt. Mater.* **2023**, *143*, 114218. [CrossRef]
9. Chen, J.M.; Jiang, Y.E.; Wang, X.; Du, L.F.; Xiao, Q.; Pan, X.; Zhou, L.; Zhou, S.L.; Peng, J.H.; Li, X.C.; et al. High-stability, high-power diode-pumped mode-locked laser with a novel Nd:Glass. *Opt. Commun.* **2024**, *558*, 130380. [CrossRef]
10. Liu, K.X.; Dong, Y.; Zhang, Z.H.; Duan, X.H.; Guo, R.H.; Zhai, Z.J.; Wang, J.L. MHz repetition rate femtosecond radially polarized vortex laser direct writing Yb:CaF_2 waveguide laser operating in continuous-wave and pulsed regimes. *Nanophotonics.* **2024**, *13*, 9–18. [CrossRef]
11. Püschel, S.; Mauerhoff, F.; Kränkel, C.; Tanaka, H. Solid-state laser cooling in Yb:CaF_2 and Yb:SrF_2 by anti-Stokes fluorescence. *Opt. Lett.* **2022**, *47*, 333–336. [CrossRef] [PubMed]
12. Zhang, G.Y.; Li, R.X.; Li, K.; Xu, H.; Zhang, B.; Niu, J.; Sui, Y.; Yuan, M.H.; Liu, X.P.; Ma, Y.J.; et al. 10-mJ 300-fs 1-kHz cryogenically cooled Yb:CaF_2 regenerative amplifier. *Opt. Commun.* **2024**, *565*, 130687. [CrossRef]
13. Zhao, C.C.; Zhang, P.X.; Li, S.M.; Fang, Q.N.; Xu, M.; Cheng, Z.Q.; Hang, Y. Development of rare-earth ion doped fluoride laser crystal. *J. Synth. Cryst.* **2022**, *51*, 1573–1587.
14. Petit, V.; Doualan, J.L.; Camy, P.; Ménard, V.; Moncorgé, R. CW and tunable laser operation of Yb^{3+} doped CaF_2. *Appl. Phys. B* **2004**, *78*, 681–684. [CrossRef]

15. Hornung, M.; Liebetrau, H.; Keppler, S.; Kessler, A.; Hellwing, M.; Schorcht, F.; Becker, G.A.; Reuter, M.; Polz, J.; Körner, J.; et al. 54 J pulses with 18 nm bandwidth from a diode-pumped chirped-pulse amplification laser system. *Opt. Lett.* **2016**, *41*, 5413–5416. [CrossRef]
16. Wentsch, K.S.; Weichelt, B.; Guenster, S.; Druon, F.; Georges, P.; Ahmed, M.A.; Graf, T. Yb:CaF_2 thin-disk laser. *Opt. Express* **2014**, *22*, 1524–1532. [CrossRef]
17. Akchurin, M.S.; Basiev, T.T.; Demidenko, A.A.; Doroshenko, M.E.; Fedorov, P.P.; Garibin, E.A.; Gusev, P.E.; Kuznetsov, S.V.; Krutov, M.A.; Mironov, I.A.; et al. CaF_2:Yb laser ceramics. *Opt. Mater.* **2013**, *35*, 444–450. [CrossRef]
18. Liu, Y.; Qin, X.P.; Gan, L.; Zhou, G.H.; Zhang, T.J.; Wang, S.W.; Chen, H.T. Preparation of sub-micron spherical Y_2O_3 particles and transparent ceramics. *J. Inorg. Mater.* **2024**, *39*, 691–696. [CrossRef]
19. Feng, S.W.; Guo, Y.C.; Sun, X.M.; Fu, J.; Li, J.Q.; Jiang, J.; Qin, H.M.; Wang, H.; Yang, Y.F. Elevating photoluminescence properties of $Y_3MgAl_3SiO_{12}$:Ce^{3+} transparent ceramics for high-power white lighting. *J. Rare Earth* **2023**, *41*, 649–657. [CrossRef]
20. Wang, D.W.; Wang, J.P.; Yuan, H.C.; Liu, Z.; Zhou, J.; Deng, J.J.; Wang, X.; Wu, B.H.; Zhang, J.; Wang, S.W. Metre-scale $Y_3Al_5O_{12}$ (YAG) transparent ceramics by vacuum reactive sintering. *J. Inorg. Mater.* **2023**, *38*, 1483–1484. [CrossRef]
21. Rakov, N.; Matias, F.; Maciel, G.S. Temperature sensing performance of Er^{3+}:Yb^{3+} co-doped CaF_2 ceramic powders using near-infrared light. *J. Rare Earth.* 2024, *in press*. [CrossRef]
22. Jin, X.H.; Dong, M.J.; Kan, Y.M.; Liang, B.; Dong, S.M. Fabrication of transparent AlON by gel casting and pressureless sintering. *J. Inorg. Mater.* **2023**, *38*, 193–198. [CrossRef]
23. Yang, C.L.; Huang, J.Q.; Huang, Q.F.; Deng, Z.H.; Wang, Y.; Li, X.Y.; Zhou, Z.H.; Chen, J.; Liu, Z.G.; Guo, W. Optical, thermal, and mechanical properties of $(Y_{1-x}Sc_x)_2O_3$ transparent ceramics. *J. Adv. Ceram.* **2022**, *11*, 901–911. [CrossRef]
24. Basiev, T.T.; Doroshenko, M.E.; Fedorov, P.P.; Konyushkin, V.A.; Kuznetsov, S.V.; Osiko, V.V.; Akchurin, M.S. Efficient laser based on CaF_2-SrF_2-YbF_3 nanoceramics. *Opt. Lett.* **2008**, *33*, 521–523. [CrossRef]
25. Jiang, Y.G.; Jiang, B.X.; Zhang, P.D.; Chen, S.L.; Gan, Q.J.; Fan, J.T.; Mao, X.J.; Jiang, N.; Su, L.B.; Li, J.; et al. Transparent Nd-doped $Ca_{1-x}Y_xF_{2+x}$ ceramics prepared by the ceramization of single crystals. *Mater. Des.* **2017**, *113*, 326–330. [CrossRef]
26. Yang, Y.; Zhou, Z.; Mei, B.; Zhang, Y.; Liu, X. Fabrication and upconversion luminescence properties of Er:SrF_2 transparent ceramics compared with Er:CaF_2. *Ceram. Int.* **2021**, *47*, 17139–17146. [CrossRef]
27. Chen, X.; Zhao, B.; Chen, N.; Cheng, J.; Dang, M.; Wang, F.; Xu, X.; Wang, H. Low temperature fired CaF_2-based microwave dielectric ceramics with enhanced microwave properties. *J. Eur. Ceram. Soc.* **2022**, *42*, 4969–4973. [CrossRef]
28. Liu, Z.D.; Shen, Q.L.; Fu, S.C.; Yang, L.T.; Chen, X.; Li, S.Y.; Cao, Y.; Liu, B.; Yu, Y.S.; Jing, Q.S.; et al. Effects of LiF sintering additive on the microstructure and mechanical properties of hot-pressed CaF_2 transparent ceramics. *Opt. Mater.* **2022**, *14*, 100147. [CrossRef]
29. Wang, P.; Huang, Z.F.; Morita, K.; Li, Q.Z.; Yang, M.J.; Zhang, S.; Goto, T.; Tu, R. Influence of spark plasma sintering conditions on microstructure, carbon contamination, and transmittance of CaF_2 ceramics. *J. Eur. Ceram. Soc.* **2022**, *42*, 245–257. [CrossRef]
30. Yang, Y.; Zhou, Z.; Mei, B.; Li, W.; Zhang, Y.; Liu, X. Energy transfer and controllable colors of upconversion emission in Er^{3+} and Dy^{3+} co-doped CaF_2 transparent ceramics. *J. Eur. Ceram. Soc.* **2021**, *41*, 7835–7844. [CrossRef]
31. Chen, L.; Mei, B.C.; Li, W.W.; Zhou, Z.W.; Yang, Y.; Zhang, Y.Q. Effect of reactive raw materials and pre-loading pressure on the microstructure and transmittance of pure CaF_2 transparent ceramics. *Mater. Chem. Phys.* **2023**, *297*, 127315. [CrossRef]
32. Kuznetsov, S.V.; Alexandrov, A.A.; Fedorov, P.P. Optical Fluoride Nanoceramics. *Inorg. Mater.* **2021**, *57*, 555–578. [CrossRef]
33. Li, W.W.; Huang, H.J.; Mei, B.C.; Wang, C.; Liu, J.; Wang, S.Z.; Jiang, D.P.; Su, L.B. Effect of Yb concentration on the microstructures, spectra, and laser performance of Yb:CaF_2 transparent ceramics. *J. Am. Ceram. Soc.* **2020**, *103*, 5787–5795. [CrossRef]
34. Sarthou, J.; Aballea, P.; Patriarche, G.; Serier-Brault, H.; Suganuma, A.; Gredin, P.; Mortier, M.; Riman, R. Wet-route synthesis and characterization of Yb:CaF_2 optical ceramics. *J. Am. Ceram. Soc.* **2016**, *99*, 1992–2000. [CrossRef]
35. Aballea, P.; Suganuma, A.; Druon, F.; Hostalrich, J.; Georges, P.; Gredin, P.; Mortier, M. Laser performance of diode-pumped Yb:CaF_2 optical ceramics synthesized using an energy-efficient process. *Optica* **2015**, *2*, 288–291. [CrossRef]
36. Li, W.W.; Jing, W.; Mei, B.C.; Zhai, P.F.; Yang, Y.; Song, J.H. Effect of NaF doping on the transparency, microstructure and spectral properties of Yb^{3+}:CaF_2 transparent ceramics. *J. Eur. Ceram. Soc.* **2020**, *40*, 4572–4577. [CrossRef]
37. Li, W.W.; Huang, H.J.; Mei, B.C.; Wang, C.; Liu, J.; Wang, S.Z.; Jiang, D.P.; Su, L.B. Fabrication, microstructure and laser performance of Yb^{3+} doped CaF_2-YF_3 transparent ceramics. *Ceram. Int.* **2020**, *46*, 19530–19536. [CrossRef]
38. Liu, X.Q.; Hao, Q.Q.; Liu, J.; Liu, D.H.; Li, W.W.; Su, L.B. Yb:CaF_2–YF_3 transparent ceramics ultrafast laser at dual gain lines. *Chin. Phys. B* **2022**, *31*, 114205. [CrossRef]
39. Wei, J.B.; Toci, G.; Pirri, A.; Patrizi, B.; Feng, Y.G.; Vannini, M.; Li, J. Fabrication and property of Yb:CaF_2 laser ceramics from co-precipitated nanopowders. *J. Inorg. Mater.* **2019**, *34*, 1341–1348. [CrossRef]
40. Huang, X.Y.; Chen, G.M.; Wei, J.B.; Liu, Z.Y.; Feng, Y.G.; Tian, F.; Xie, T.F.; Li, J. Fabrication of Yb,La:CaF_2 transparent ceramics by air pre-sintering with hot isostatic pressing. *Opt. Mater.* **2021**, *116*, 111108. [CrossRef]
41. Liu, Z.Y.; Wei, J.B.; Toci, G.; Pirri, A.; Patrizi, B.; Feng, Y.G.; Xie, T.F.; Hreniak, D.; Vannini, M.; Li, J. Microstructure and laser emission of Yb:CaF_2 transparent ceramics fabricated by air pre-sintering and hot isostatic pressing. *Opt. Mater.* **2022**, *129*, 112540. [CrossRef]
42. Kitajima, S.; Yamakado, K.; Shirakawa, A.; Ueda, K.I.; Ezura, Y.; Ishizawa, H. Yb^{3+}-doped CaF_2-LaF_3 ceramics laser. *Opt. Lett.* **2017**, *42*, 1724–1727. [CrossRef] [PubMed]

43. Guo, L.H.; Shi, Y.; Tian, F.; Chen, H.H.; Toci, G.; Pirri, A.; Patrizi, B.; Vannini, M.; Li, J. Microstructure and laser performance of fine-grained Yb:CaF_2 transparent ceramics prepared by two-step sintering. *Opt. Mater.* **2023**, *140*, 113841. [CrossRef]
44. Li, Q.; Wang, Y.; Wang, J.; Ma, J.; Ni, M.; Lin, H.; Zhang, J.; Liu, P.; Xu, X.D.; Tang, D.Y. High transparency Pr:Y_2O_3 ceramics: A promising gain medium for red emission solid-state lasers. *J. Adv. Ceram.* **2022**, *11*, 874–881. [CrossRef]
45. Li, X.; Yin, J.; Lai, Y.M.; Zhang, X.; Yu, S.Q. Improved microstructure and optical properties of Nd:YAG ceramics by hot isostatic pressing. *Ceram. Int.* **2023**, *49*, 31939–31947. [CrossRef]
46. Zhang, L.X.; Li, X.Y.; Hu, D.J.; Liu, Z.Y.; Xie, T.F.; Wu, L.X.; Yang, Z.X.; Li, J. Fabrication and properties of non-stoichiometric $Tb_2(Hf_{1-x}Tb_x)_2O_{7-x}$ magneto-optical ceramics. *J. Adv. Ceram.* **2022**, *11*, 784–793. [CrossRef]
47. Li, W.W.; Huang, H.J.; Mei, B.C.; Song, J.H. Comparison of commercial and synthesized CaF_2 powders for preparing transparent ceramics. *Ceram. Int.* **2017**, *43*, 10403–10409. [CrossRef]
48. Akinribide, O.J.; Mekgwe, G.N.; Akinwamide, S.O.; Gamaoun, F.; Abeykoon, C.; Johnson, O.T.; Olubambi, P.A. A review on optical properties and application of transparent ceramics. *J. Mater. Res. Technol.* **2022**, *21*, 712–738. [CrossRef]
49. Liu, Z.D.; Ji, Y.M.; Xu, C.Y.; Wang, Y.; Liu, Y.; Shen, Q.; Yi, G.; Yu, Y.; Mei, B.; Liu, P.; et al. Microstructural, spectroscopic and mechanical properties of hot-pressed Er:SrF_2 transparent ceramics. *J. Eur. Ceram. Soc.* **2021**, *41*, 4907–4914. [CrossRef]
50. Yi, G.Q.; Liu, Z.D.; Li, W.W.; Mei, B.C.; Yin, S.M.; Xue, L.H.; Yan, Y.W. Gd^{3+} doping induced microstructural evolution and enhanced visible luminescence of Pr^{3+} activated calcium fluoride transparent ceramics. *Ceram. Int.* **2023**, *49*, 7333–7340. [CrossRef]
51. Chen, J.M.; Mei, B.C.; Li, W.W.; Zhang, Y.Q. Fabrication and spectral performance of Nd_x:$(La_{0.05}Lu_{0.05}Gd_{0.05}Y_{0.05})Ca_{0.8-x}F_{2.2+x}$ High-entropy transparent fluoride ceramics. *Ceram. Int.* **2024**, *50*, 6128–6134. [CrossRef]

ceramics

MDPI

Article

Improving the Transparency of a $MgAl_2O_4$ Spinel Damaged by Sandblasting through a SiO_2-ZrO_2 Coating

Akram Zegadi [1,2,*], Abdelwahhab Ayadi [2,3], Ikram Khellaf [2,4], Mohamed Hamidouche [1,2], Gilbert Fantozzi [5], Alicia Durán [6] and Yolanda Castro [6]

1 Emerging Materials Research Unit, Ferhat Abbas University Setif 1, Setif 19000, Algeria; mhamidouche@univ-setif.dz
2 Optics and Precision Mechanics Institute, Ferhat Abbas University Setif 1, Setif 19000, Algeria; abdelwahhab.ayadi@univ-setif.dz (A.A.); khellaf_ikram@univ-setif.dz (I.K.)
3 Laboratory of Non-Metallic Materials, Optics and Precision Mechanics Institute, Ferhat Abbas University Setif 1, Setif 19000, Algeria
4 Laboratory of Applied Optics, Institute of Optics and Precision Mechanics, University Ferhat Abbas Setif 1, Setif 19000, Algeria
5 Institut National des Sciences Appliquées Lyon, Université Claude Bernard Lyon 1, CNRS, MATEIS, UMR5510, 69621 Villeurbanne, France; gilbert.fantozzi@insa-lyon.fr
6 Instituto de Cerámica y Vidrio (CSIC), Campus de Cantoblanco, 28049 Madrid, Spain; aduran@icv.csic.es (A.D.); castro@icv.csic.es (Y.C.)
* Correspondence: a.zegadi@univ-setif.dz or akzegadi@gmail.com

Abstract: Transparent materials in contact with harmful environments such as sandstorms are exposed to surface damage. Transparent $MgAl_2O_4$ spinel used as protective window, lens or laser exit port, among others, is one of the materials affected by natural aggressions. The impact of sand particles can cause significant defects on the exposed surface, thus affecting its optical and mechanical behavior. The aim of this work is to improve the surface state of a spinel damaged surface by the deposition of a thin layer of SiO_2-ZrO_2. For this purpose, spinel samples obtained from different commercial powders sintered by Spark Plasma Sintering were sandblasted and further coated with a SiO_2-ZrO_2 thin layer. The coating was successfully synthesized by the sol/gel method, deposited on the sandblasted samples and then treated at 900 °C, reaching a final thickness of 250 nm. The results indicated that sandblasting significantly affects the surface of the spinel samples as well as the optical transmission, confirmed by UV-visible spectroscopy and profilometry tests. However, the deposition of a SiO_2-ZrO_2 coating modifies the UV-visible response. Thus, the optical transmission of the S25CRX12 sample presents the best transmission values of 81%, followed by the S25CRX14 sample then the S30CR sample at 550 nm wavelength. An important difference was observed between sandblasted samples and coated samples at low and high wavelengths. At low wavelengths (around 200 nm), sandblasting tends to improve significantly the transmission of spinel samples, which exhibit a low transmission in the pristine state. This phenomenon can be attributed to the healing of small superficial defects responsible for the degradation of transmission such as pores or flaws. When the initial transmission at 200 nm is high, the sandblasting worsens the transmission. Sandblasting reduces slightly the transmission values for long wavelengths due to the formation of large superficial defects like chipping by creation and propagation of lateral cracks. The coating of the sandblasted samples exhibits some healing of defects induced by sandblasting. The deposition of the SiO_2-ZrO_2 layer induces a clear increase in the optical transmission values, sometimes exceeding the initial values of the transmission in the pristine state.

Keywords: spinel; $MgAl_2O_4$; SiO_2-ZrO_2 coatings; sandblasting; SPS sintering; optical transmission

1. Introduction

The degradation of optical and mechanical properties of materials used for precise imaging and detection of IR or near-IR radiation produced by particles suspended in the air, such as silica particles in the Saharan regions, is a serious problem [1,2]. Transparent ceramics are very promising materials in the field of protection of IR detection devices thanks to their high transparency in the range between 300 and 900 nm wavelength and their interesting mechanical properties [3,4]. However, these materials suffer erosion–corrosion as a consequence of the exposition of natural attacks such as sandstorms. Studies based on the fragility of materials such as glass are numerous, but in the case of transparent ceramics, they are limited. In fact, numerous works of research [5,6] have shown that erosion of brittle materials is affected by many factors, including the properties of the incident particles (properties of the target materials (i.e., their hardness, fracture toughness and their surface condition) and the test conditions (i.e., impact speed, impact angle and temperature). During sandstorms, all these factors take place at the same time and in a random way (wide range of grain size, variable shape of sand particles, variable speeds during the same storm, variable impact angles, etc.), producing defects in the surface. Indeed, the cracks and defects created act as effective scattering centers for the incident radiation. Therefore, optical transmission and resolution are affected, decreasing the performance of detection equipment. Cracking also weakens the material and may be enough to cause failure of the protective window [3].

Magnesium aluminate spinel ($MgAl_2O_4$) is one of the most important ceramic materials used as transparent shielding ceramics and also as protective windows for sensor applications in aeronautics since these applications require high optical and mechanical performances [7–9]. Spinel possesses very good mechanical properties, i.e., hardness, toughness and wear resistance, high transmission and an appropriate refractive index [10,11]. For possible applications in aggressive environments, these characteristics should be preserved for a long time to maintain the structural integrity and optical transparency. Mechanical properties and transparency depend on powder characteristics (particle size and chemical purity) and subsequent sintering conditions. Only a few research studies have addressed the surface degradation of transparent spinel. For example, Lallemant et al. [2] studied the erosion of several types of transparent materials including spinel. They observed that the increase in surface roughness caused by erosion directly affected the optical quality of spinel. They concluded that the variation depends on three parameters: wavelength, microstructure and hardness. Thus, if the size of the defects (roughness Rq) is small compared to the wavelength, the reflectance of the surface increases, and, consequently, the optical properties also increase, except for the infrared, where the variation can be neglected. Tokariev et al. [8] reported that coarse-grained spinel wear resists erosion better than fine-grained spinel. On the other hand, Von Helden et al. [4] studied the behavior of spinel under abrasion by performing erosion tests using sand particles at different durations. They found that spinel exhibits minimal degradation of its surface state after erosion and that hardness has a direct influence on short-term wear of the samples. However, the stress intensity factor KIC influenced long-term wear.

To improve the wear resistance of spinel, different groups of researchers have considered two different approaches: (1) to strengthen the mechanical properties of the surface through thermal or thermochemical treatments before exposure to sandstorms and (2) the deposition of thin transparent layers on the surface of the damaged material to restore the properties (optical transmission and mechanical strength) [12]. Various techniques are used to prepare these layers, such as chemical vapor deposition (CVD), physical vapor deposition (PVD) or atomic layer deposition (ALD), among others. However, the sol-gel technique is a good alternative to obtain adherent and homogeneous coatings on a wide variety of substrates, including $MgAl_2O_4$ spinel, with good stoichiometric control of the synthesis process [13,14]. Several researchers have reported that the deposition of SiO_2, ZrO_2 and SiO_2–ZrO_2 coatings on soda-lime glass significantly increases the optical transmission and mechanical strength of the glass [15,16]. Zhang et al. [17] reported that the deposition of SiO_2-ZrO_2 coatings on glass improves the wear properties and controls the degradation. Zhang et al. [18] also studied the indentation properties of SiO_2-ZrO_2 coatings on glass substrates and observed that increasing the ZrO_2 amount leads to a remarkable increase in hardness. A hardness of 22 GPa and a Young's modulus of 193 GPa were achieved for 100% ZrO_2. In general, the deposition of a sol-gel coating can improve the mechanical properties by healing the superficial defects [18], filling flaws or crack tips [19]. The incorporation of ZrO_2 in the composition of the coatings is beneficial because ZrO_2 offers mechanical resistance, good surface properties and good biocompatibility and biological properties [17,20], making it a beneficial material for reinforcement. To our knowledge, there has been no study focused on the deposition of a SiO_2-ZrO_2 coating on spinel except the paper of DiGiovanni et al. [3]. In this paper, the authors consider that the success of the method is limited, particularly at high temperatures, and thus, more studies are necessary to consider the feasibility of the method.

In this work, spinel samples fabricated from different spinel powders by Spark Plasma Sintering (SPS) at different temperatures are sandblasted to simulate erosion by sand particles. Then, a SiO_2-ZrO_2 coating is deposited on the eroded spinel samples and the surface roughness and optical transmission are analyzed and compared with pristine spinel samples.

2. Experimental Procedure

2.1. Materials and Methods

2.1.1. Powders and SPS Sintering

$MgAl_2O_4$ spinel is produced from pure powders supplied by Baikowski (La Balme de Sillingy, France) in three different batches (S25XRX12, S25CRX14, S30CR). Table 1 shows the chemical composition, specific surface area (SSA) and crystallite size of each batch. The S30CR powder has the highest sulfur content (600 ppm) compared to S25CRX12 (200 ppm) and S25CRX14 (300 ppm). The commercial powders were placed into a graphite matrix and then sintered by SPS (HPD 25, FCT System) without any further preparation. A pressure of 72 MPa was applied at ambient temperature. The heating rate was controlled at 100 °C/min up to above 800 °C. Then, the temperature was increased up to 1100 °C with a heating rate of 10 °C/min. At these stages, the increase of punch displacement can be mainly attributed to the powder shrinkage. Maximum densification is reached after heating to the final sintering temperature with a low heating rate (1 °C/min), without holding time. By using a low heating rate, residual porosity and pore size can be reduced without significant grain growth. At the end of the heating up to the final sintering temperature, the pressure is removed. The sintering cycle is completed by a subsequent annealing at 1000 °C for 10 min to eliminate any residual stresses. The cooling rate between the end of sintering and annealing and after subsequent annealing is 100 °C/min. The used sintering cycle was previously optimized to obtain transparent samples of $MgAl_2O_4$ spinel [21]. Three sintering temperatures, 1290 °C, 1310 °C and 1330 °C, were chosen to obtain different microstructures. The resulting samples were polished on both surfaces following a well-defined polishing protocol [9] in order to obtain the best possible surface conditions. The final thickness of the samples varies between 1.9 and 2.4 mm.

Table 1. Average chemical composition (in ppm by weight) of the used powders (for the S25CRX12 and S25CRX14 powders, the abbreviated names S 12 and S 14 are used).

	Chemical Composition Weight (ppm)						Crystallite Size (nm)	SSA (m^2/g)
	Na	K	Fe	Si	Ca	S		
S 12	11	14	8.1	13	4.3	200	29	17.1
S 14	11	13	6.5	14	6.9	300	60	27.4
S30CR	13	35	1	36	<1	600	73	31

2.1.2. Optical Transmission

The samples were characterized using a UV-visible spectrophotometer (Shimadzu UV-1800, Shimadzu corporation, Kyoto, Japan) to follow the evolution of optical transmission before and after sandblasting and after deposition of the layer. The spectral band was recorded between 200 nm and 1100 nm. Since thicknesses of the samples are different, the transmittance measurement was normalized at a thickness of 0.88 mm using Equation (1) [21]:

$$T(d_2) = (1 - R_S)\left(\frac{T(d_1)}{1 - R_S}\right)^{d_2/d_1} \quad (1)$$

T (d_1) = transmittance for a sample of starting thickness d_1
T (d_2) = transmittance for a sample of thickness d_2 = 0.88 mm
RS = total normal surface reflectance (0.14)

2.1.3. Microscopic Observations

Scanning electron microscopy (SEM) was used to analyze the morphology of the different spinel samples obtained by SPS. SEM images were obtained on Zeiss Supra 55VP (Zeiss, Oberkochen, Germany) and analyzed by the free software ImageJ (version 1.50b) using a three-dimensional correction factor of 1.225 [22].

Figure 1 shows the microstructural morphology of the spinel samples sintered at 1310 °C using the different powders. S25CRX14 has a higher porosity than the other ones, associated with the lower compressibility of the powder [9]. The fine pores are essentially located at the multiple junctions of the grains, and their average size is around 100 nm (Figure 1). As reported in reference [9], a small amount of residual porosity can significantly weaken the transparency. The grain size was also determined from the microscopic images. Table 2 summarizes the grain size for the different spinel samples sintered at different temperatures. It is worth noting that increasing the sintering temperature results in the formation of larger grains for all the used powders. The S30CR powder sample sintered at 1330 °C shows the highest grain size compared to the others.

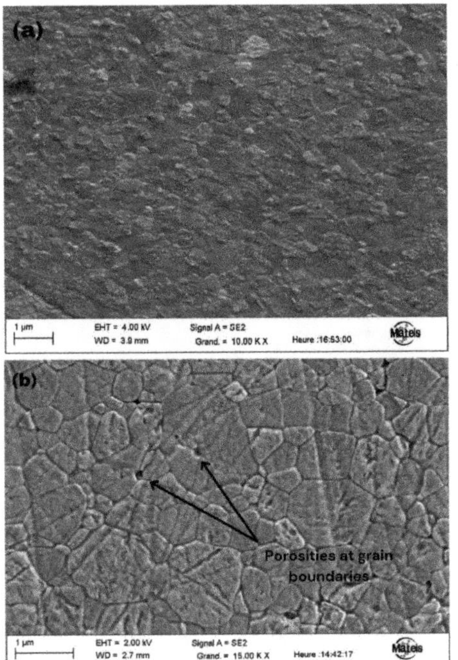

Figure 1. *Cont.*

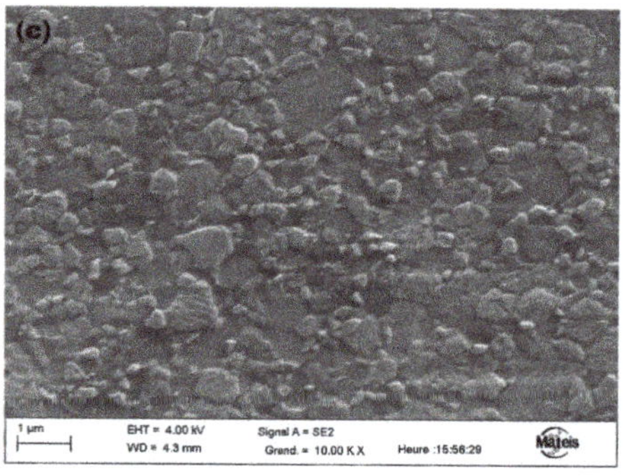

Figure 1. SEM images of samples (**a**) S25CRX12, (**b**) S25CRX14 and (**c**) S30CR sintered at 1310 °C.

Table 2. Grain size (nm) for the S25CRX12, S25CRX14 and S30CR samples sintered at 1290 °C, 1310 °C and 1330 °C by SPS.

	Grain Size (nm)		
Sintering temperature (°C)	1290	1310	1330
S 12	338	500	835
S 14	366	628	805
S30CR	408	614	964

2.1.4. Sandblasting

Sandblasting of the sintered spinel samples was carried out using a horizontal sandblasting device according to the standards for erosion tests by particles suspended in air (DIN 50 332 [23] and ASTM G76 [24]). The flow speed was set at 30 m/s, and the angle of incidence was 90°. According to previous research, the 90° angle is considered the most harmful angle for fragile materials such as alumina during particle erosion [25]. Finally, the projected mass used was 200 g. The sand used in this study comes from the desert region of Ouargla (southern Algeria) and was used as it is, without any preliminary washing. The sand grains come in various shapes (round and elliptical) and different colors, as shown in Figure 2. This suggests that the chemical composition of the sand is not pure, and it may contain silica (transparent) and silica mixed with other metal oxides. Sand microhardness, determined in a previous study [26] (under a load of 0.6 N for a batch of 30 particles), is 14.49 ± 3.28 GPa.

Figure 2. Micrograph of used sand particles from the desert region of Ouargla.

The sand particles of the desert region of Ouargla have wide particle size distribution. For that, using several sieves of different sizes and a vibratory sieve shaker, a sorting was carried out to keep only the particles of size between 250 and 850 μm (Table 3). The particle size distribution of the sand used is determined by a laser particle size analyzer, and the results are presented in Figure 3.

Table 3. Sand particle size distribution parameters.

Diameter (D)	Particles Size (μm)
D (0.1) minimum diameter	235
D (0.5) average diameter	515
D (0.9) maximum diameter	815

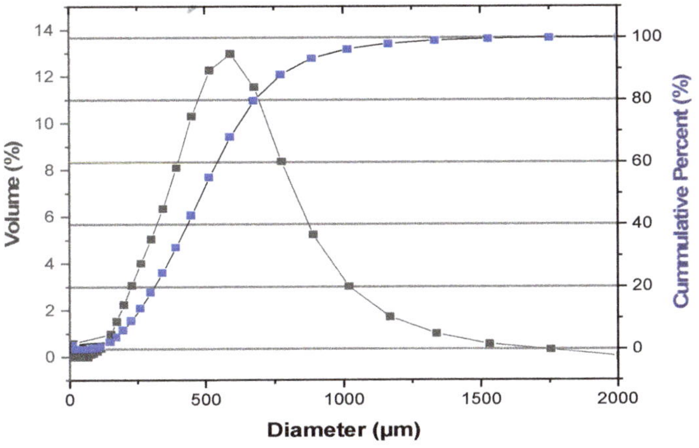

Figure 3. Particle size distribution of the used sand.

Then, the samples were sandblasted and characterized. The surface roughness of the central area of the sample was characterized before and after sandblasting with a profilometry instrument (Form Talysurf Series 120i by Taylor Hobson Company, Leicester, UK). The Ra roughness parameter (Equation (2)) is universally recognized, and it is the most used roughness parameter, defined as the arithmetic average of the absolute values of the Z(x) ordinates within a base length. As well, Z(x) is the function which describes the profile of the analyzed surface in terms of height (Z) and position (x) of the sample over an evaluation length "l".

$$R_a = \int_0^{l_0} |Z_i(x)| dx \quad (2)$$

The morphology of the coated surface was observed by atomic force microscopy (Asylum MFP3D, Oxford Instruments Asylum Research, Santa Barbara, CA, USA).

2.1.5. Coating of Sintered Samples

Silica-zirconia (SiO_2-ZrO_2) films were prepared by the hydrolysis and condensation of tetraethoxysilane (TEOS, $Si(OC_2H_5)_4$, Sigma-Aldrich, Darmstadt, Germany, 99%), methyltriethoxysilane (MTES, $CH_3Si(OC_2H_5)_3$) (Sigma-Aldrich, 98%) and zirconium n-butoxide (ZrBu) as precursors. The sol was prepared in two steps; TEOS and MTES were first pre-hydrolyzed with acidified water (0.1 N HCl) for 2 hours at 40 °C. The molar ratios used were MTES/TEOS = 20/80. In a second step, ZrBu was mixed with acetylacetone in a molar ratio of ZrBu/AcAc = 1 and stirred for one hour. Then, distilled water was added while maintaining agitation for 1 h at 25 °C. Once both solutions were obtained, they were slowly mixed, and after 15 minutes, the remaining water was added dropwise (H_2O/(TEOS + MTES + ZrBu) = 1.1) until the completion of hydrolysis, and the mixture was kept for 2 hours at 40 °C. The molar composition used is SiO_2/ZrO_2 = 95/05. The final concentration of (SiO_2-ZrO_2) was fixed to 130 g/L.

SiO_2-ZrO_2 sol was analyzed by an infrared spectrophotometer to monitor the hydrolysis and condensation reactions that occurred during the solution preparation. Figure 4 presents the FTIR transmittance spectra in the range of 4000–400 cm^{-1} of the silica-zirconia sol. The three dominant peaks characteristic of the Si-O were mentioned: The band at near to 1070 cm^{-1} is attributed to antisymmetric stretching of the oxygen atoms in the bond of Si-O-Si. The frequency peak below 450 cm^{-1} is attributed to the oscillatory movements of oxygen atoms perpendicular to the Si-O-Si plane or deformation vibrations of O-Si-O. The band at frequency near 878 cm^{-1} is associated with the symmetric stretching motion of oxygen atoms. These peaks confirm the hydrolysis reaction of the silica precursors and the formation of the Si-O-Si bond. The bands at 1262 cm^{-1} and 2900–3000 cm^{-1} are assigned to the Si-CH_3 bonds.

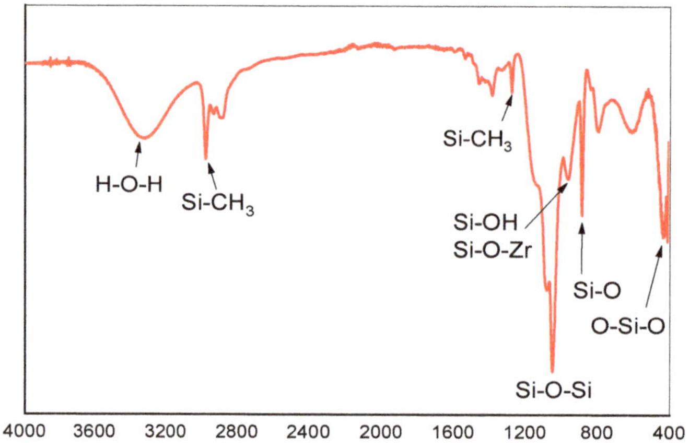

Figure 4. Infrared spectrum of the SiO_2-ZrO_2 solution.

The various bands observed in the infrared spectroscopy curve are summarized in Table 4.

Table 4. Different structures in the SiO_2-ZrO_2 solution depending on wave number.

N°	Wave Number (cm^{-1})	Vibration	Structure	References
1	3323	Stretching O-H and Si-O-H	H-O-H/H_2O	[27]
2	2973	ν_s C-H	Si-CH_3/-CH_3	[28]
3	1262	δ_s C-H	Si-CH_3	[29]
4	1042	ν_{as} Si-O-Si	Si-O-Si	[30]
5	940	ν_β Si-O	≡Si-OH Si-O-Zr	[31] [32]
6	878	ν_β Si-O	Si-O	[33]
7	432	δ O-Si-O	O-Si-O	[30]

Vibration types: ν_s symmetric stretching vibration, ν_{as} antisymmetric stretching vibration, ν_β in-plane stretching vibration, δ deformation vibration, δs symmetric deformation vibration (bending).

According to Gonçalves et al. [32], the band located at around 940 cm^{-1} can be attributed to Si-O-Zr or Si-OH. The emergence of this band indicates the incorporation of zirconia into the silica network.

It is important to note that the solution stability is crucial to prevent poor deposition on the sample, thus the viscosity of the solution was measured using a vibro-viscosimeter SV-1A (A&D Company, Tokyo, Japan), and a value of 3. 71 mPa·s was obtained.

SiO_2-ZrO_2 sol was deposited on transparent spinel substrates by the dip-coating method. The samples were carefully cleaned with water and detergent and rinsed with distilled water and finally with ethanol in an ultrasonic bath, followed by drying. The deposition layer was carried out at a constant withdrawal rate of 3 cm/min.

After deposition, the obtained samples were heat treated in a furnace at 900 °C for 1 h to eliminate organic residues, densify the layer and impart final properties to the deposited layer. Typically, treatment of layers deposited on fragile materials such as glasses is carried out at temperatures below 500 °C. However, in our case, the layer was treated at 900 °C, which is likely to provide the deposited layer with very interesting optical and especially mechanical properties. This treatment helps to reduce the porosity by eliminating open pores and reducing pore size [34]. The deposited layer has an average thickness of 250 nm, measured by a mechanical profilometry.

3. Results and Discussion

The spinel samples before and after being sandblasted and then after the deposition of SiO_2-ZrO_2 coatings were characterized by a UV-visible spectrophotometer. Figures 5–7 show the UV-vis spectra between 200 and 1100 nm for the S25CRX12, S25CRX14 and S30CR spinel samples sintered at different temperatures before and after sandblasting and after the deposition of a SiO_2-ZrO_2 coating.

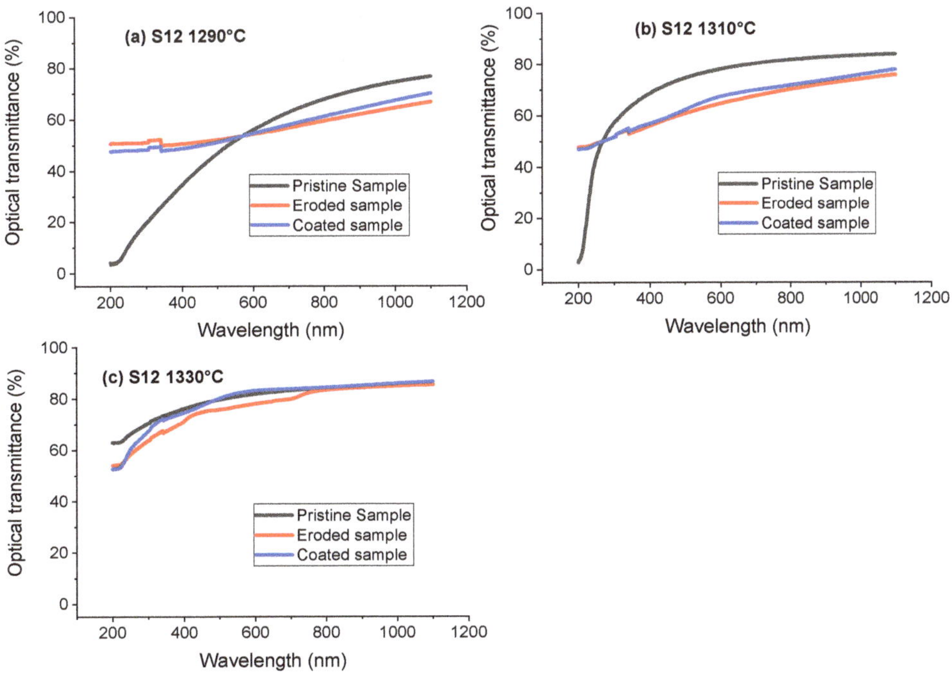

Figure 5. Evolution of optical transmission before and after sandblasting, then after coating of S25CRX12 spinel samples sintered at 1290, 1310 and 1330 °C.

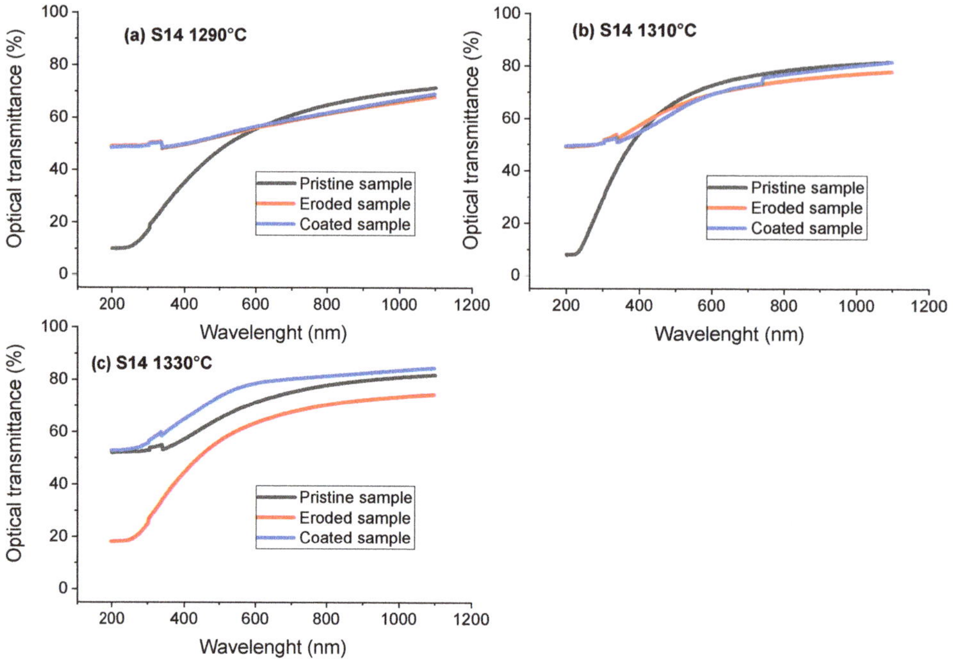

Figure 6. Evolution of optical transmission before and after sandblasting then after coating of S25CRX14 spinel samples sintered at 1290, 1310 and 1330 °C.

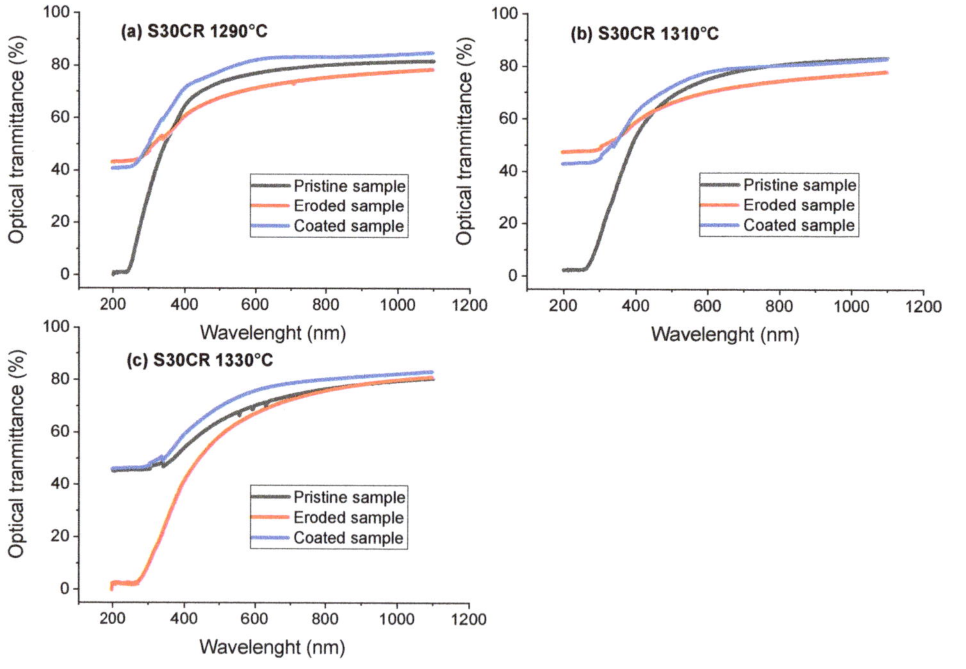

Figure 7. Evolution of optical transmission before and after sandblasting then after coating of S30CR spinel samples sintered at 1290, 1310 and 1330 °C.

Figure 5 shows the evolution of optical transmission for S25CRX12 samples sintered at 1290, 1310 and 1330 °C. The optical transmission depends on the sintering temperature. In the pristine state at 200 nm, samples sintered at lower temperatures exhibit very low optical transmission, whereas the transmittance becomes important after sintering at 1330 °C. The increase in sintering temperature leads to an increase in optical transmission at 550 nm wavelength. It varies from 51.8 to 80.9% when the sintering temperature varies from 1290 to 1330 °C. At 1000 nm, similar effects were observed.

Concerning the damaged samples, at 200 nm, it is important to note that sandblasting increases the optical transmission of the fabricated spinel samples by approximately 40%, excepting for the sample sintered at 1330 °C, where sandblasting did not produce a great effect. Sandblasting has no significant effect on the optical transmission at 550 nm wavelength for the sintered samples, whatever the sintering temperature. At 1000 nm, sandblasting also degraded slightly the optical transmission of the samples, except for the sintering at 1330 °C.

After deposition of the coating layer, very small changes in optical properties are observed for 200, 500 and 1000 nm wavelengths; just a slight increase in the transmittance can be noted.

Figure 6 shows the UV-vis spectra for the S25CRX14 samples sintered at 1290, 1310 and 1330 °C before and after sandblasting and after the deposition of SiO_2-ZrO_2 coating. It is obvious that for the pristine state, increasing the sintering temperature increases the optical transmission of the spinel samples in the visible range at 550 nm wavelength. At 200 nm wavelength, the optical transmission for the pristine state is very low for the sintering at 1290 and 1310 °C, and for the sintering at 1330 °C, a relatively higher optical transmittance is noted. At 1000 nm wavelength, there is also an increase in optical transmission while increasing the sintering temperature.

Sandblasting of the spinel samples has an effect on the optical transmission of the S25CRX14 samples. Indeed, at 550 nm wavelength, sandblasting has slightly damaged the surface reflected by the reduction in the optical transmission values from 70 to 67 and from 69 to 60% for the sintering temperatures 1310 and 1330 °C, respectively. At 200 nm wavelength, the optical transmission is increased from 10 to 49%, and from 8 to 49% for the sintering temperatures 1290 and 1310 °C, respectively. For the sintering at 1330 °C, the optical transmission is reduced from 52 to 18%. Then, at 1000 nm wavelength, sandblasting reduces slightly the optical transmission for all the samples.

After deposition of the layer on the spinel samples, a clear improvement in the optical transmission at 550 nm wavelength of the sample sintered at 1310 °C is observed. At 550 nm wavelength, it can be noted that the coating has little effect on the transmission for the sintering at 1290 and 1310 °C, and an improvement in the optical transmission is observed for the sample sintered at 1330 °C. We also notice that at 200 nm wavelength, by applying the coating, an improvement in the optical transmission of the sandblasted sample sintered at 1330 °C is achieved from 18 to 53%. On the other hand, the coating has little effect on samples sintered at 1290 and 1310 °C. At 1000 nm wavelength, we can consider that the coating improves optical transmission, especially for samples sintered at 1310 and 1330 °C. The transmission varies from 73% after sandblasting to 84% after coating for the sample sintered at 1330 °C.

Figure 7 represents the variation of the optical transmission as a function of the wavelength of the S30CR samples sintered at 1290, 1310 and 1330 °C, respectively. In the pristine state, the samples exhibit at 200 nm wavelength a low optical transmission for the two sintering temperatures 1290 °C and 1310 °C and a relatively high transmission for the sintering at 1330 °C. At 1000 nm wavelength, all samples have high transmittance values up to 83%.

For the 200 nm wavelength, sandblasting has a considerable effect on the recorded values. For the sintering temperatures 1290 °C and 1310 °C, sandblasting increases significantly the optical transmission, whereas for the sintering temperature 1330 °C, the opposite effect is observed, with a significant reduction in the transmission. At 550 and

1000 nm wavelengths, sandblasting affected optical transmission by reducing slightly the optical properties.

Finally, after coating of the samples, we note a clear improvement in the optical transmission at 200 nm wavelength for the samples sintered at 1330 °C and a small effect for the sintering temperatures 1290 °C and 1310 °C. At 550 and 1000 nm, the coating increases slightly the transmission for all the sintering temperatures. Also, it is interesting to note that the transmission values after coating even exceed the initial values in the pristine state (see Tables 5–7).

Table 5. Optical transmission values recorded at 200 nm for the S25CRX12, S25CRX14 and S30CR samples sintered at 1290, 1310 and 1330 °C.

	Transmittance at λ = 200 nm								
Sintering Temperature	**Pristine (%)**			**Eroded (%)**			**Coated (%)**		
	S 12	**S 14**	**S30CR**	**S 12**	**S 14**	**S30CR**	**S 12**	**S 14**	**S30CR**
1290 °C	4.2	10	0	50.7	49.0	43.1	47.6	48.5	40.8
1310 °C	2.8	8.1	2.47	47.8	49.1	47.4	47.0	49.5	42.9
1330 °C	62.9	52.1	45.34	54.1	18.2	0	52.6	52.8	46.1

Table 6. Optical transmission values recorded at 550 nm for the S25CRX12, S25CRX14 and S30CR samples sintered at 1290, 1310 and 1330 °C.

	Transmittance at λ = 550 nm								
Sintering Temperature	**Pristine (%)**			**Eroded (%)**			**Coated (%)**		
	S 12	**S 14**	**S30CR**	**S 12**	**S 14**	**S30CR**	**S 12**	**S 14**	**S30CR**
1290 °C	51.8	52.1	75.5	53.2	54.3	69.5	52.9	54.9	80.2
1310 °C	76.6	70.2	72.6	62.9	67.4	68.4	65.2	66.6	75.6
1330 °C	80.9	68.5	67.6	76.9	60.2	63.2	82.1	76.5	73.1

Table 7. Optical transmission values recorded at 1000 nm for the S25CRX12, S25CRX14 and S30CR samples sintered at 1290, 1310 and 1330 °C.

	Transmittance at λ = 1000 nm								
Sintering Temperature	**Pristine (%)**			**Eroded (%)**			**Coated (%)**		
	S 12	**S 14**	**S30CR**	**S 12**	**S 14**	**S30CR**	**S 12**	**S 14**	**S30CR**
1290 °C	74.6	69.7	81.4	64.5	65.9	77.6	67.4	66.2	84.3
1310 °C	83.5	81.1	82.7	74.2	77.3	77.0	75.9	80.4	82.1
1330 °C	85.2	80.9	79.5	84.9	73.4	79.9	86	83.5	82.3

The optical transmission values recorded at 200, 550 and 1000 nm for the S25CRX12, S25CRX14 and S30CR samples sintered at 1290, 1310 and 1330 °C are summarized in Tables 5–7.

From the results, it is possible to conclude that the transmission in the visible range depends significantly on the sintering temperature. The optical transmission presents the best values in the pristine state for the S25CRX12 powder, followed by the S30CR and the S25CRX14. Grain size measurements indicate that all samples present a small grain size ranging from approximately 300 to 900 nm (Table 2). It can also be noted that increasing the sintering temperature leads to the formation of larger grains for all the powders. Therefore, it appears that the good optical transmission depends neither on grain size nor on the surface area of the starting powders because the best performing sample, S25CRX12, has the lowest specific surface area (17.1 m^2/g) and a relatively large grain size

(835 nm). These interesting results can be attributed to the significant higher purity of this powder (S25CRX12) and, in particular, to the low sulfur content mentioned previously (Table 1). S25CRX14 contains a pore volume greater than the other two powders, with this powder exhibiting a lower compressibility [9]. The fine pores are especially located near the multiple grain junctions with a mean size of about 100 nm.

It can be noted that at 200 nm wavelength, the sandblasting tends to improve the transmission of spinel samples when the pristine transmission is low, which corresponds to some healing of small defects responsible for transmission, such as pores or flaws. Studies have shown that sensitivity to small defects is greater at low wavelengths. On the other hand, when the initial optical transmission is high for short wavelengths, the sandblasting of samples provokes a significant decrease in the transmission. For high wavelengths, the effect of sandblasting on the degradation of the optical transmission is lower. This can be attributed to the formation of large defects like chipping, as indicated by several studies [35] which report that after sandblasting transparent samples such as glasses or transparent ceramics, material removal takes place by formation and propagation of lateral cracks which develop into chipping. Evenly, increasing surface reflection can also be a contributing factor to the decrease in optical transmission [26]. Table 8 presents the roughness values Ra of the spinel samples in the pristine, sandblasted and then coated states.

Table 8. Roughness Ra of spinel samples S25CRX12, S25CRX14 and S30CR before and after sandblasting and after coating.

	Ra (nm)								
Sintering Temperature	**Pristine**			**Eroded**			**Coated**		
	S 12	**S 14**	**S30CR**	**S 12**	**S 14**	**S30CR**	**S 12**	**S 14**	**S30CR**
1290 °C	6.3	4.6	8.6	15	14.9	27.6	12	6.7	5.7
1310 °C	8.3	7.6	10.1	25.3	16.7	19.2	6.7	8.6	7
1330 °C	6.4	5.5	12	23.1	39.2	22.5	8.8	8.3	8.1

Sandblasting degrades the surface of the samples, starting with a roughness Ra of around 8 nm in the pristine state to a higher roughness around 25 nm after sandblasting, which clearly affects the optical transmission of the material in most spectra, although, at 200 nm, there is a healing phenomenon, and the transmission increases. The transmission increases after sandblasting when the initial transmission is low, indicating a healing phenomenon.

The coating increases the optical transmission after sandblasting for all the wavelengths. This increase is very significant at 200 nm when the transmission after sandblasting is very low. The roughness values decrease after the deposition of the coating. On S30CR samples sintered at 1290, 1310 and 1330 °C and the S25CRX12 sample sintered at 1310 °C, the roughness Ra has lower values than in the pristine state, proving the effectiveness of the deposited layer in restoring the characteristics of an eroded material.

On S25CRX12 powder sintered at 1330 °C, and at a low wavelength of 200 nm, the coating improves the optical transmission lost during sandblasting. However, if sandblasting increases transmission (for the sintering temperatures 1290 and 1310 °C), the coating effect becomes very small to negligible. For the same wavelengths and for the S25CRX14 and S30CR powders, we have a behavior similar to that just shown. Then, at visible wavelengths, i.e., 550 nm, the S25CRX12 powder exhibits surface degradation, which implies a reduction in the optical transmission of the samples sintered at 1310 and 1330 °C. The same effect is observed on the S25CRX14 sample sintered at 1330 °C; otherwise, on the other samples, the effect of the coating is almost unremarkable. In the case of S30CR powder, all samples, regardless of their sintering temperature, are affected by sandblasting. Then, their optical properties are restored after coating. We can say that by applying a silica-based coating, the initial characteristics of the eroded spinel samples are largely restored, namely roughness and optical transmission. Indeed, sol-gel coatings can penetrate into the micro-defects and defects generated during sandblasting by filling them, adding an improvement in the morphology of the non-eroded zone. Previous results have shown that the coating heals defects in the pristine surface, thereby decreasing light reflection [26]. To better explain the restoration of surface roughness, an atomic force microscopy study was carried out. Figure 8 shows the variation in the roughness profile after sandblasting for the S25CRX14 and S30CR samples and then after coating.

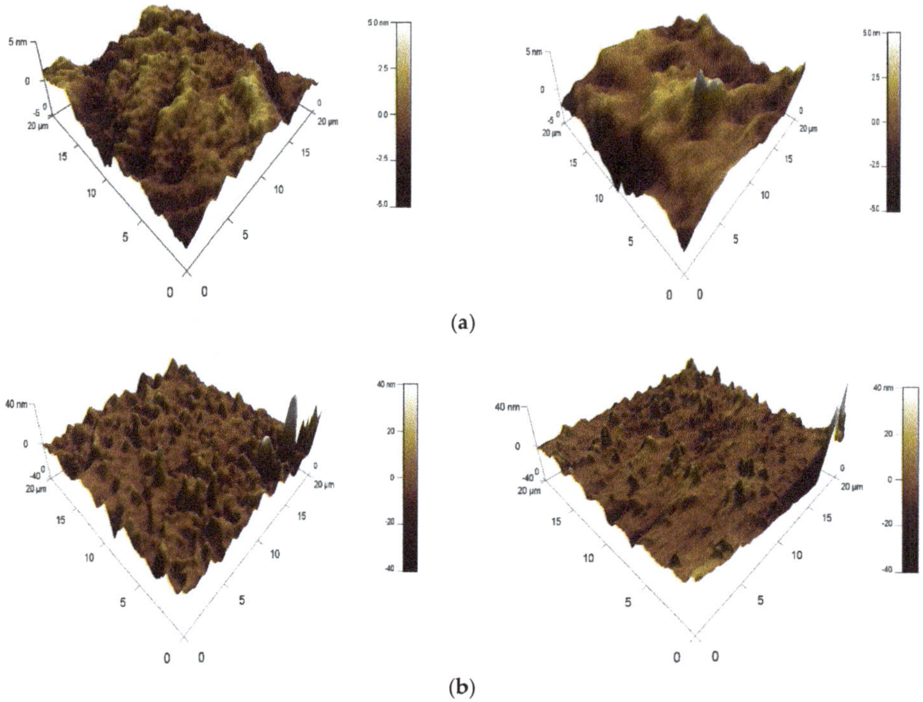

Figure 8. Surface roughness after sandblasting and after coating for samples; left: uncoated, right: coated. (**a**) S25CRX14 sintered at 1310 °C, (**b**) S30CR sintered at 1310 °C.

From the microscopic images, it is clear that the surface roughness degraded by sandblasting is much improved after coating. This results in the filling of defects resulting from sandblasting, leading to a reduction in reflection and consequently an increase in optical transmission. Thus, the coating can restore optical properties and even exceed the transmission values recorded for the pristine sample. This is observed on samples of S25CRX 12 powders sintered at 1330 °C, S25CRX14 sintered at 1330 °C and on S30CR sintered at 1290 °C (Figure 9). This behavior can be linked to a decrease in the effect of sharp-bump defects, which became more spherical, as observed by Ayadi et al. [26] for glasses. It should be useful to make some SEM observations in order to specify this point. On the other hand, roughness alone is not sufficient to explain the results. Sandblasting can also increase the optical transmission of a spinel sample at short wavelengths thanks to a phenomenon of a defect's healing such as pores or flaws.

At long wavelengths (1000 nm), the sintered S25CRX12 powder shows a degradation of the optical properties after sandblasting then a restoration of these properties after coating, except for the samples where sandblasting has no effect on the pristine state; then, a minimal effect can be cited. The same phenomenon was recorded on the S25CRX14 and S30CR powders, where there is no effect of the coating when sandblasting does not degrade the surface or optical transmission. On the other hand, when the sandblasting provokes a strong reduction in optical transmission, a significant improvement after coating is observed.

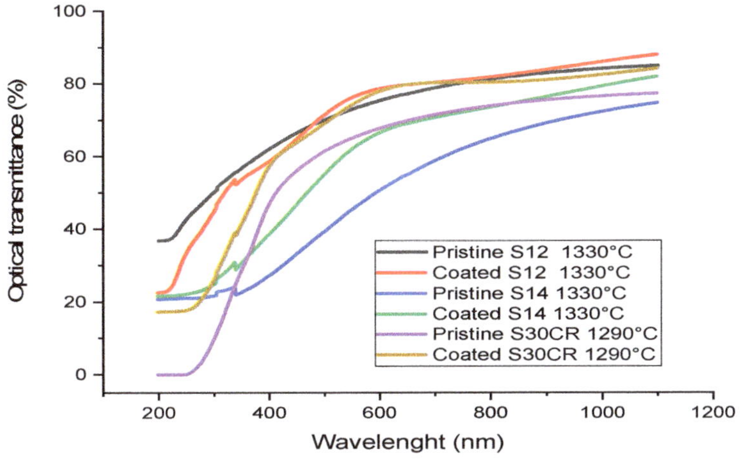

Figure 9. Best optical transmission values obtained after coating.

For the future, it would be interesting to investigate the damage caused on the surfaces eroded by sandblasting and the coating effect by making some microscopic observations with an optical or a scanning electronic microscope. Furthermore, it would be suitable to consider the two other alternative scenarios:

- A sandblasted and coated sample is again exposed to sandblasting; thus, the optical endurance of the repaired ceramic can be determined.
- A pristine sample is directly coated and characterized without sandblasting: what is the potential of such treatment and how it will perform when sandblasting.

4. Conclusions

Sand erosion of a transparent spinel fabricated with different powders (S25CRX12, S25CRX14, S30CR) sintered at different temperatures (1290, 1310, 1330 °C) was carried out, simulating the harsh conditions of a sandstorm in the Sahara. Then, a SiO_2-ZrO_2 layer deposition was carried out after sandblasting in order to correct the damage of the eroded surface. The results obtained allowed us to conclude with the following points:

- In the pristine state, the optical transmission presents the best transmission values (85% at 1000 nm) for the S25CRX12 powder, followed by the S30CR then the S25CRX14. This good result can be attributed to the significant difference which lies in the high purity of this powder, particularly the sulfur content.
- At 200 nm wavelength, sandblasting tends to improve the transmission of spinel samples (nearly 40%) for the specimens exhibiting a low transmission in the pristine state. This behavior can be attributed to the healing of small superficial defects responsible for the degradation of transmission, such as pores or flaws, for instance. Evenly, when the pristine transmission at 200 nm is high, the sandblasting worsens the transmission, which can be due to the formation of impact flaws.
- Sandblasting reduces slightly the transmission values for long wavelengths due to the formation of large superficial defects like chipping by creation and propagation of lateral cracks. Furthermore, sandblasting degrades the surface of the samples, starting with a roughness Ra of around 8 nm in the pristine state to a significant roughness after sandblasting around 25 nm.
- Using a silica-based coating can largely restore the initial roughness and optical transmission of the eroded spinel samples by repairing the superficial defects and

filling flaws or cracks tips. After coating, the optical transmission is restored, and sometimes it is higher than the initial value.

Author Contributions: Conceptualization, A.Z. and A.A.; methodology, A.Z., A.A. and G.F.; validation, A.Z., A.A., M.H., A.D., Y.C. and G.F.; formal analysis, A.Z., M.H., A.D., Y.C. and G.F.; writing—original draft preparation, A.Z., A.A., I.K. and G.F.; writing—review and editing, A.Z., A.A., I.K., A. D., Y.C. and G.F.; visualization, A.Z., A.A., I.K. and G.F.; supervision, A.Z., M.H. and G.F. All authors have read and agreed to the published version of the manuscript.

Funding: The research received no external funding.

Data Availability Statement: The original contributions presented in the study are included in the article, further inquiries can be directed to the corresponding author.

Acknowledgments: The authors are grateful to L. Bonneau and A. Vivet from Baikowski (La Balme de Sillingy, France) for the supplying of the spinel powders.

Conflicts of Interest: The authors declare no conflicts of interest. Baikowski, the supplier of the spinel powders, had no role in the design, execution, interpretation, or writing of the study.

References

1. Telling, R.H.; Jilbert, G.H.; Field, J.E. Erosion of aerospace materials by solid-particle impact. In *Proceedings Volume 3060, Window and Dome Technologies and Materials V*; SPIE: Bellingham, WA, USA, 1997. [CrossRef]
2. Lallemant, L.; Garnier, V.; Bonnefont, G.; Marouani, A.; Fantozzi, G.; Bouaouadja, N. Effect of solid particle impact on light transmission of transparent ceramics: Role of the microstructure. *Opt. Mater.* **2014**, *37*, 352–357. [CrossRef]
3. DiGiovanni Anthony, A.; Fehrenbacher, L.; Roy, D.W. Hard transparent domes and windows from magnesium aluminate spinel. In *Proceedings Volume 5786, Window and Dome Technologies and Materials IX*; SPIE: Bellingham, WA, USA, 2005. [CrossRef]
4. Von Helden, S.; Malzbender, J.; Krüger, M. Mechanical properties, wear resistance and surface damage of glasses and $MgAl_2O_4$ spinel ceramic after abrasion and scratch exposure. *Ceram. Int.* **2019**, *45*, 10765–10775. [CrossRef]
5. Lindsley, B.A.; Marder, A.R. The effect of velocity on the solid particle erosion rate of alloys. *Wear* **1999**, *225–229*, 510–516. [CrossRef]
6. Wellman, R.G.; Allen, C. The effects of angle of impact and material properties on the erosion rates of ceramics. *Wear* **1995**, *186–187*, 117–122. [CrossRef]
7. Krell, A.; Klimke, J.; Hutzler, T. Advanced spinel and sub-µm Al_2O_3 for transparent armor applications. *J. Eur. Ceram. Soc.* **2009**, *29*, 275–281. [CrossRef]
8. Tokariev, O.; Schnetter, L.; Beck, T.; Malzbender, J. Grain size effect on the mechanical properties of transparent spinel ceramics. *J. Eur. Ceram. Soc.* **2013**, *33*, 749–757. [CrossRef]
9. Zegadi, A.; Kolli, M.; Hamidouche, M.; Fantozzi, G. Transparent $MgAl_2O_4$ spinel fabricated by spark plasma sintering from commercial powders. *Ceram. Int.* **2018**, *44*, 18828–18835. [CrossRef]
10. Swab, J.J.; Pavlacka, R.; Gilde, G.; Kilczewski, S.; Wright, J.; Harris, D. Determining the Strength of Coarse-Grained ALON and Spinel. *J. Am. Ceram. Soc.* **2014**, *97*, 592–600. [CrossRef]
11. Tustison, R.W. *Window and Dome Technologies and Materials XI. Orlando, Florida, United States. Proceedings of SPIE, 7302*; SPIE: Bellingham, WA, USA, 2009.
12. Pellice, S.; Gilabert, U.; Solier, C.; Castro, Y.; Durán, A. Mechanical behavior of glass reinforced with SiO_2 hybrid sol–gel coatings. *J. Non-Cryst. Solids* **2004**, *348*, 172–179. [CrossRef]
13. Gallardo, J.; Durán, A.; Di Martino, D.; Almeida, R.M. Structure of inorganic and hybrid SiO_2 sol–gel coatings studied by variable incidence infrared spectroscopy. *J. Non-Cryst. Solids* **2002**, *298*, 219–225. [CrossRef]
14. Innocenzi, P.; Abdirashid, M.O.; Guglielmi, M. Structure and properties of sol-gel coatings from methyltriethoxysilane and tetraethoxysilane. *J. Sol-Gel Sci. Technol.* **1994**, *3*, 47–55. [CrossRef]
15. García-Heras, M.; Rincón, J.M.; Romero, M.; Villegas, M.A. Indentation properties of ZrO_2–SiO_2 coatings on glass substrates. *Mater. Res. Bull.* **2003**, *38*, 1635–1644. [CrossRef]
16. Endres, K.; Mennig, M.; Amlung, M.; Gier, A.; Schmidt, H. Enhancement of fracture strength of cutted plate glass by the application of SiO_2 sol-gel coatings. *Thin Sol. Films* **1999**, *351*, 132–136. [CrossRef]
17. Zhang, Q.; Li, X.; Shen, J.; Wu, G.; Wang, J.; Chen, L. ZrO_2 thin films and ZrO_2-SiO_2 optical reflection filters deposited by sol–gel method. *Mat. Lett.* **2000**, *45*, 311–314. [CrossRef]
18. Zhang, H.; Zhou, J.-H.; Zhang, Q.-L.; Yang, H. Mechanical and Optical Properties of Ion-exchange Strengthened Glass Coated with Sol-Gel Derived ZrO_2-SiO_2 Film. *J. Inorg. Mater.* **2013**, *28*, 779–785. [CrossRef]
19. Ma, K.J.; Chien, H.H.; Huang, S.W.; Chen, S.C.; Chao, C.L. Crack Filling of Cover Glasses by Sol-Gel Coatings. *Adv. Mater.* **2013**, *797*, 700–705. [CrossRef]

20. Nogami, M. Glass preparation of the ZrO_2-SiO_2 system by the sol-gel process from metal alkoxides. *J. Non-Crys Solids* **1985**, *69*, 415–423. [CrossRef]
21. Bonnefont, G.; Fantozzi, G.; Trombert, S.; Bonneau, L. Fine-grained transparent $MgAl_2O_4$ spinel obtained by spark plasma sintering of commercially available nano powders. *Ceram. Int.* **2012**, *38*, 131–140. [CrossRef]
22. Apetz, R.; Van Bruggen, M.P.B. Transparent Alumina: A Light-Scattering Model. *J. Am. Ceram. Soc.* **2003**, *86*, 480–486. [CrossRef]
23. *DIN 50332*; Strahlverschleiβprufung (Ray of Schleiβ Test). DIN: Berlin, Germany, 1984.
24. *ASTM G76-04*; Standard Test Method for Conducting Erosion Tests by Solid Particle Impingement Using Gas Jets. ASTM International: West Conshohocken, PA, USA, 2004.
25. Oka, Y.I.; Mihara, S.; Yoshida, T. Impact-angle dependence and estimation of erosion damage to ceramic materials caused by solid particle impact. *Wear* **2009**, *267*, 129–135. [CrossRef]
26. Ayadi, A.; Bouaouadja, N.; Durán, A.; Castro, Y. Hybrid SiO_2–ZrO_2 coatings for restoring and repairing glasses damaged by sandblasting. *Ceram. Int.* **2020**, *46*, 10634–10640. [CrossRef]
27. Brinker, C.J.; Scherer, G.W. *Sol-Gel Science, The Physics and Chemistry of Sol-Gel Processing;* Academic Press: Cambridge, MA, USA, 1990; 463p, ISBN 13:978-0-12-134970-7.
28. Al-Oweini, R.; El-Rassy, H. Synthesis and characterization by FTIR spectroscopy of silica aerogels prepared using several $Si(OR)_4$ and $R''Si(OR')_3$ precursors. *J. Mol. Struct* **2009**, *919*, 140–145. [CrossRef]
29. Gopal, N.O.; Narasimhulu, K.V.; Rao, J.L. EPR, optical, infrared and Raman spectral studies of Actinolite mineral. *Spectrochim. Acta. Mol. Biomol. Spectrosc.* **2004**, *60*, 2441–2448. [CrossRef] [PubMed]
30. Bertoluzza, A.; Fagnano, C.; Antonietta Morelli, M.; Gottardi, V.; Guglielmi, M. Raman and infrared spectra on silica gel evolving toward glass. *J. Non-Cryst. Solids* **1982**, *48*, 117–128. [CrossRef]
31. Socrates, G. *Infrared and Raman Characteristic Group Frequencies: Tables and Charts*, 3rd ed.; Wiley: Abingdon, UK, 2004; 368p, ISBN 978-0-470-09307-8.
32. Gonçalves, R.R.; Guimarães, J.J.; Ferrari, J.L.; Maia, L.J.Q.; Ribeiro, S.J.L. Active planar waveguides based on sol–gel Er^{3+}-doped SiO_2–ZrO_2 for photonic applications: Morphological, structural and optical properties. *J. Non-Cryst. Solids* **2008**, *354*, 4846–4851. [CrossRef]
33. Chmel, A. Vibrational spectra and defect structure of silica prepared by non-organic sol-gel process. *J. Non-Cryst. Solids* **1990**, *122*, 285–290. [CrossRef]
34. Sakka, S. *Handbook of Sol-Gel Science and Technology: Processing, Characterization and Applications, Volume I: Sol-Gel Processing;* Kluwer Academic Publishers: Boston, MA, USA; Springer Science and Business Media: New York, NY, USA, 2005; 680p, ISBN 1402079664/9781402079665.
35. Ćurković, L.; Kumić, I.; Grilec, K. Solid particle erosion behavior of high purity alumina ceramics. *Ceram. Int.* **2011**, *37*, 29–35. [CrossRef]

ceramics

MDPI

Article

Enhancing Transparency in Non-Cubic Calcium Phosphate Ceramics: Effect of Starting Powder, LiF Doping, and Spark Plasma Sintering Parameters

Kacper Albin Prokop [1,2], Sandrine Cottrino [3], Vincent Garnier [3], Gilbert Fantozzi [3], Yannick Guyot [4], Georges Boulon [4] and Małgorzata Guzik [1,2,*]

1 Faculty of Chemistry, University of Wrocław, ul. F. Joliot-Curie 14, 50-383 Wrocław, Poland; kacper2311@gmail.com
2 Łukasiewicz Research Network—PORT Polish Center for Technology Development, ul. 147 Stabłowicka, 54-066 Wroclaw, Poland
3 Institut National des Sciences Appliquées Lyon, Universite Claude Bernard Lyon 1, CNRS, MATEIS, UMR5510, 69621 Villeurbanne, France; sandrine.cottrino@insa-lyon.fr (S.C.); gilbert.fantozzi@insa-lyon.fr (G.F.)
4 Institut Lumière Matière, Universite Claude Bernard Lyon 1, CNRS, UMR5306, 69100 Villeurbanne, France; yannick.guyot@univ-lyon1.fr (Y.G.); georges.boulon@univ-lyon1.fr (G.B.)
* Correspondence: goguzik@poczta.fm; Tel.: +48-71-37-51-373

Abstract: Our objective is to achieve a new good-quality and mechanically durable high-transparency material that, when activated by rare earth ions, can be used as laser sources, scintillators, or phosphors. The best functional transparent ceramics are formed from high-symmetry systems, mainly cubic. Considering hexagonal hydroxyapatite, which shows anisotropy, the particle size of the initial powder is extremely important and should be of the order of several tens of nanometers. In this work, transparent micro-crystalline ceramics of non-cubic $Ca_{10}(PO_4)_6(OH)_2$ calcium phosphate were fabricated *via* Spark Plasma Sintering (SPS) from two types of nanopowders i.e., commercially available (COM. HA) and laboratory-made (LAB. HA) *via* the hydrothermal (HT) protocol. Our study centered on examining how the quality of sintered bodies is affected by the following parameters: the addition of LiF sintering agent, the temperature during the SPS process, and the quality of the starting nanopowders. The phase purity, microstructure, and optical transmittance of the ceramics were investigated to determine suitable sintering conditions. The best optical ceramics were obtained from LAB. HA nanopowder with the addition of 0.25 wt.% of LiF sintered at 1000 °C and 1050 °C.

Keywords: non-cubic calcium phosphates; nano-crystalline powders; SPS; transparency; optical material

1. Introduction

Over the past two decades, transparent ceramics have gained significant attention as advanced optical ceramic materials. Recently, the list of applications of transparent ceramics, for which some of them are highly sophisticated, for laser media, phosphors, scintillators, armor windows, infrared domes, and electro-optical components has widely increased in all domains and has impacted our daily lives [1–3]. They are recognized for their combination of optical transparency and outstanding mechanical properties. Moreover, their application in photonics, particularly with rare earth doping, has become prominent.

It is now well known that this is not universally achievable for all inorganic compounds and is significantly impacted by factors such as light scattering due to residual pores and the presence of secondary phases or impurities. Also, potential candidates for well-light-transmitting ceramics have to crystallize in the high-symmetry system (preferably in cubic). This group of materials is very limited, and taking into account non-doped and RE^{3+}-doped host lattices, it contains only garnets ($Y_3Al_5O_{12}$, $Lu_3Al_5O_{12}$), sesquioxides (Y_2O_3, Sc_2O_3, and Lu_2O_3), spinels ($MgAl_2O_4$), fluorides (CaF_2), selenides (ZnSe), sulfides (ZnS), and perovskite-type BMT ($Ba(MgZrTa)O_3$), primarily utilized in lasers, scintillating materials, or solid-state lighting, as previously highlighted in our earlier publications.

For polycrystalline non-cubic ceramics, additional scattering losses due to birefringent splitting of the beam at grain boundaries occur. This scattering depends on the grain size and is smaller the smaller the grain size is (the grain size must be in the sub-μm range) [4].

Given the extensive research conducted on transparent ceramics, it was revealed that the incorporation of MgO, LiF, or SiO_2 as sintering aids can improve the transparency of sintered bodies [5–7]. For example, the groundbreaking fabrication of transparent ceramic Nd-doped yttrium aluminum garnet, Nd:YAG, was performed by Ikesue using SiO_2 as the sintering aid [8]. This cubic transparent ceramics has been demonstrated to be superior to its single-crystal counterpart for laser applications [9]. Nonetheless, certain transparent ceramics, such as $MgAl_2O_4$ spinel [10], Al_2O_3 [11,12], CaF_2 [13], and $(La,Y)_2O_3$ [14], can be crafted without employing a sintering aid. This is frequently achieved through a synergistic combination of appropriate initial powders and a densification process that imparts a substantial driving force to the sintering process, namely via techniques like Hot Isostatic Pressing (HIP), Hot Pressing (HP), or Spark Plasma Sintering (SPS) [15].

One of our research programs is focused on exploring novel materials (mainly cubic) with the potential to yield highly transparent optical ceramics from families other than those already known and applied. For example, our studies comprised cubic molybdates of Y_6MoO_{12} [16], molybdato-tungstates of La_2MoWO_9 [17–19], tungstates (articles under review), and most recently, eulytite-type phosphates of $Ba/Sr_3Y(PO_4)_3$ [20], from which we have succeeded in obtaining some translucent ceramic bodies.

Current research involves different interesting phosphates characterized by high thermal and chemical stability, such as calcium phosphate. As it crystallizes in a hexagonal structure, the size and homogeneity of the initial powder are extremely important. The fabrication of transparent ceramics from non-cubic raw materials is possible, but their nanoparticles should be of the order of several tens of nanometers.

In the literature, one can find a few papers reporting studies on polycrystalline transparent ceramics based on phosphates from three groups: $Ca_3(PO_4)_2$ β-TCP [21], hydroxyapatite $Ca_{10}(PO_4)_6(OH)_2$ (HA) [22], and fluorapatite $Ca_{10}(PO_4)_3F_2$ (FAP) [23]. They crystallize in the trigonal (s. g., *R3c*, no. 111) and hexagonal (*P63/m*, no. 176) crystal systems, respectively. Among them, calcium phosphate transparent ceramics were considered a good candidate for storage phosphors in dosimetry and may also be applied as scintillators for α-ray detection [24], while fabricating transparent bio-compatible hydroxyapatite ceramics with nanosized grains is important for direct observations of in vivo interactions with proteins and/or cells [25,26].

In turn, the most important achievement for fluorapatites is a diode-pumped anisotropic (non-cubic) ceramic laser that uses microdomain-controlled Nd^{3+}-doped hexagonal fluorapatite (Nd^{3+}:$Ca_{10}(PO_4)_6F_2$, Nd:FAP) polycrystalline ceramics as the gain medium. They were fabricated by the RE^{3+}-assisted magnetic grain-orientation control method as a step toward achieving giant micro-photonics [27,28]. Furuse et al. demonstrated laser oscillation in randomly oriented non-cubic FAP ceramics activated with Nd^{3+} and Yb^{3+} ions [29,30].

All these transparent materials were obtained *via* advanced sintering techniques, i.e., SPS. Eriksson et al. conducted intriguing research, exploring the use of a high-pressure SPS approach to achieve transparency in SPS processes conducted at lower temperatures but under elevated pressure conditions [31]. However, none of these studies used sintering aids.

In this work, we chose non-cubic calcium phosphate as $Ca_{10}(PO_4)_6(OH)_2$ hydroxyapatite to investigate the effect of using different amounts of LiF on the quality of the obtained sintered body. LiF is a well-known sintering aid with a melting point of approximately 850 °C [2,3]. When utilized as an additive in the densification processes of ceramic materials, particularly in methods involving uniaxial pressure such as HP and SPS, LiF plays a crucial role. It manifests a distinctive capacity to generate a low-viscosity lubricating film upon compact particles. This film, in turn, facilitates the processes of particle sliding and rearrangement, thereby creating enhanced densification through the mechanism of liquid-phase sintering. The use of LiF was particularly useful to obtain transparent ceramic materials of $MgAl_2O_4$ spinel [32], but also YAG [33], Dy^{3+}:Y_2O_3 [34], Sm^{3+}:Y_2O_3 [35], Nd^{3+}:Lu_2O_3 [36], and MgO [37].

The use of sintering additives may reduce the temperature, which is optimal for obtaining a transparent material; therefore, the effect of the temperature applied during the sintering on the quality of the sintered bodies was also checked.

From the articles of Kim et al. and Kato et al. [22,24], we know that the fabrication of transparent ceramics based on calcium phosphate is possible. However, the experiments described by the Japanese researchers used a commercially available powder from Japan, which is not accessible to us. As in the next stages of our project, we want to activate hydroxyapatite with rare earth ions, so optimizing the sintering conditions to obtain the highest quality transparent undoped host lattice was crucial. This is why we also checked the effect of the starting powder used for the process on the optical quality of our sintered specimens. All sintering procedures were performed via the SPS process, which is an efficient method of consolidating polycrystalline materials at relatively low temperatures in a short time.

2. Materials and Methods

2.1. Powders Used for Sintering via SPS Method

Two types of nanopowders, i.e., commercially available and synthesized by us, were used for sintering experiments. The commercially available powder was tricalcium phosphate $Ca_3(PO_4)_2 * \times H_2O$ (Sigma-Aldrich, St. Louis, MI, USA) with particle size <200 nm (BET), which was in reality hexagonal hydroxyapatite, as we will show in the next part. So, we named it in this paper COM. HA. The second nano-crystalline powder was obtained in our laboratory.

Here, it is necessary to mention that the literature is rich in reports on hydroxyapatite synthesis methods, particularly those employing hydrothermal [38,39] or precipitation reactions [40] as well as sol–gel techniques [41–43]. Microwave radiation has also been explored as a novel means to achieve higher purity and ultrafine-sized powders [44,45]. Additionally, innovative synthesis protocols, such as mechano-chemical synthesis [46] and surfactant-modified hydrothermal methods [47], have been explored.

We decided to fabricate nanopowdered hydroxyapatite *via* hydrothermal (HT) reaction following the protocol described in [48]. This laboratory-made powder is named in the paper as LAB. HA. For this synthesis, commercially available $Ca(NO_3)_2 * 4H_2O$ (Alfa Aesar, 99.98%) and analytically pure $NH_4H_2PO_4$ (Chempur, 99.5%) were used. Proper quantities of calcium nitrate and ammonium phosphate were individually dissolved in distilled water. The controlled addition of the $NH_4H_2PO_4$ solution to the $Ca(NO_3)_2$ solution, executed drop-wise with continuous stirring, led to the formation of a precipitate, achieving a targeted Ca/P ratio of 1.67. The initial pH of the suspension was approximately 5.1, but as a result of the gradual introduction of NH_4OH (Sigma-Aldrich, ACS reagent grade), the pH increased to achieve the desired value of 11.0. The addition of NH_4OH not only makes it possible to precise pH control, but it also serves to protect against carbonate formation during the synthesis process. After vigorous stirring for an additional 10 min, the solution was transferred into a Teflon-lined HT reactor and heated at 200 °C for 24 h. Under natural cooling to RT, the particulate product was collected through centrifugation. The collected powder underwent a thorough washing process involving repeated resuspension in distilled water and subsequent particle sedimentation through centrifugation at 6000 rpm for 5 min. This washing procedure was repeated five times, with a concluding methanol rinse to reduce feasible agglomeration in the final dried powder. The drying process was executed in an oven at 80 °C for 12 h.

2.2. Sintering by SPS

The samples were sintered using an SPS sintering machine, type FCT System HP D25, Rauenstein, Germany. The nanopowder of calcium phosphate was poured into a graphite die with an internal diameter of 10 mm. The die was surrounded with a 5 mm thick layer of graphite wool to ensure thermal insulation. The interior surface of the graphite die was lined with a 0.35 mm thick graphite foil. Temperature monitoring was facilitated by an axial pyrometer focused on a designated orifice positioned 3 mm above the material surface within the die. Pellets were produced under vacuum conditions, maintaining consistent powder quantities (0.5 g) inserted between the pistons in the die for all experiments. In some cases, the appropriate amounts of 0.25 to 1 wt.% of LiF (Alfa Aesar, 99.98%) were added to the initial powder and ground in an agate mortar using a small amount of ethanol. The sintering conditions used in numerous tests were similar in terms of pressure, moment of pressure application, temperature rise, and dwell time. However, the sintering temperature was modified. Initially, the temperature was increased from 20 to 750 °C for 30 min. under pressure of 38 MPa and kept at this temperature for 10 min. Subsequently, the temperature was raised from 750 °C to the expected temperature (950–1100 °C), respectively, while the applied axial pressure increased to 100 MPa. The temperature was held for 15 min while maintaining the applied pressure. Sintering was performed in a vacuum. Finally, the pressure was reduced to 38 MPa and the temperature to 950 °C in 10 min. Then the specimen was cooled to RT naturally. The sintering cycle of calcium phosphate powder was chosen based on the studies of Kim et al. [22] and. Kato et al. [24]. The sample thickness after sintering was in the range of 1.03–1.36 mm. Before characterization to eliminate any geometric imperfections and remove the carbon-contaminated layer resulting from contact with the graphite die and punch, both sides of the sintered pellets were polished, as described below. All samples were polished to the same thickness of approximately 1 mm.

2.3. Analysis and Characterization Techniques

Structural analysis and Rietveld refinements from X-ray powder diffraction data. The powder X-ray diffraction patterns for nanopowdered samples and micro-ceramics were recorded at RT using a Bruker D8 Advance X-ray diffractometer with Ni-filtered $CuK_{\alpha1+2}$ radiation ($K_{\alpha1+2}$, λ = 1.5418 Å). The step rate was 0.032° per step, and the counting time was 1.5 s per step within the range of 10–80° 2θ. The diffractograms were compared with the simulated XRD patterns of hexagonal $Ca_{10}(PO_4)_6(OH)_2$ (PDF 01-086-1203) and trigonal $Ca_3(PO_4)_2$ (PDF 00-70-2065), respectively. X-ray diffraction analyses were realized to check if the desired initial phases are present in the powder and that there is no phase decomposition during sintering. Also, it was checked whether the obtained specimens did not contain a trace of contamination by carbon originating from the graphite die.

The Rietveld refinements were performed for sintered bodies in order to determine the ratio between the two phases, i.e., hexagonal and trigonal, present in each sintered body. Calculations were performed using the HighScore Plus 3.0 software, with the initial parameters derived from crystal data of two phases: hexagonal $Ca_{10}(PO_4)_6(OH)_2$ s. g., *P63/m* no. 176 (COD #1011242) and trigonal $Ca_3(PO_4)_3$ s. g., *R3c*, no. 161 (COD #1517238). A determined background was used, and peak shapes were modeled using a pseudo-Voigt function. Global variables, specifically specimen displacement [mm], were refined for the obtained patterns. Structural parameters involved the scale factor and unit cell parameters. Peak profile variables were fine-tuned using "U left", "V left", and "W left" with respect to Peak Shape 1 Left. A crucial parameter for achieving well-refined results was the incorporation of preferred grain orientation during calculations. Due to the distinct crystal cell parameters (a and c) of the two phases, the intensities of XRD diffraction lines vary. Refining this parameter resulted in achieving low percentages of R_p and R_{wp} values (where R_p is the R profile and R_{wp} is the weighed profile), hence more precise phase ratios for the analyzed samples.

2.3.1. Microscopic Analysis of Starting Powders by TEM

To check the quality, particle size, and morphology of nanopowders used for sintering *via* SPS, the Transmission Electron Microscopy (TEM) technique was employed. The measurements were carried out on a H-8100 TEM of a Hitachi company (Tokyo, Japan) equipped with a Penta FET EDX detector (Oxford Instruments, UK). An acceleration voltage of 200 kV was applied. The nanopowders were suspended in ethanol, subsequently deposited onto a copper grid coated with carbon, and then dried in air.

2.3.2. Microscopic Analysis of Sintered Ceramics by SEM

All sintered bodies were investigated by scanning electron microscopy (SEM) using a Hitachi S-3400 N equipped with an energy-dispersive X-ray spectroscopy EDS detector, Thermo Scientific Ultra Dry. The observations were made in the secondary electron (SE) mode with a high voltage of 2.00 kV and a beamed current equal to 50 pA. The ceramic samples were not coated with any gold alloy layer; however, to perform these measurements, they were properly polished and annealed for 1h at a temperature 100 °C lower than the sintering one.

2.3.3. Disc Polishing

The polishing process proceeded in the following manner: first, diamond shields with grit sizes of 200, 500, 1200, and 4000 μm were employed. Then the diamond grains, suspended in a lubricating fluid, were applied to the polishing cloth with grit sizes of 5 μm, 3 μm, 1 μm, and 100 nm. The polishing effect was continuously controlled after each polishing cloth to obtain the best-quality mirror-faced polished samples.

2.3.4. Density Measurements

The density of the sintered samples was measured using Archimedes' method with distilled water. The values of the relative densities were calculated assuming a theoretical density of 3.13 g/cm^3 for the hexagonal phase and 3.14 g/cm^3 for the trigonal phase, and taking into account the ratio of both phases for each ceramics. The measurement uncertainty was ± 0.2.

2.3.5. Transmission Measurements of Sintered Ceramics

To evaluate the optical quality of the obtained polished ceramics, the total optical transmission of the specimens was measured at RT in the spectral region of 300–850 nm using an Edinburgh FLS980 spectrofluorometer, equipped with a 450 W xenon continuous arc lamp, R928P, R2658P PMT detectors, and an integrating sphere. The transmission was normalized to 1 mm thickness.

3. Results

3.1. Raw Nanopowder Characterizations

Before sintering, both powders, i.e., COM. HA and LAB. HA, were carefully investigated in terms of phase purity (XRD) and morphology (TEM).

3.1.1. Phase Analysis

The hexagonal crystal system is the most commonly occurring structure for hydroxyapatite. It is characterized by *P63/m* (no. 176) space group symmetry and lattice parameters of $a = b = 9.432$ Å, $c = 6.881$ Å, and $\gamma = 120°$. The structure comprises an arrangement of PO_4 tetrahedra, which are interconnected by Ca^{2+} ions scattered throughout. These Ca^{2+} are sited in two distinct crystal positions: in precisely aligned columns (Ca(I)) and within equilateral triangles (Ca(II)) that revolve around the screw axis. The OH groups are situated in columns on the screw axes, whereby neighboring OHs diverge in rival directions. It is important to note that such an arrangement suggests the existence of steric hindrance among the closest OHs. Hence, to inverse the direction of OH^- within a column, it is necessary to eliminate some of the OH^- to create reversal points. It may be achieved by substituting OH^- with a vacancy, F^-, Cl^-, or any suitable replacement [49].

β-TCP is a high-temperature phase that is typically obtained by thermal conversion of amorphous calcium phosphate and has the most stable tricalcium phosphate crystalline structure. This material was confirmed to have a rhombohedral structure (s. g., *R3c*, no. 161). Unit-cell parameters with higher precision ($a = b = 10.4352(2)$ Å, $c = 37.4029(5)$ Å, $\alpha = \beta = 90°$, and $\gamma = 120°$ in the hexagonal setting) and positional parameters for oxygen with equal precision were obtained by the neutron powder diffraction technique, compared with the single-crystal X-ray diffraction data by Dickens et al. [50]. The site Ca(4) with atomic coordinates (0.0, 0.0, −0.0851(6)) was confirmed to be very different from the other four Ca sites. The position Ca(4) is three-fold coordinated with oxygen atoms and has a lower occupancy factor of 0.43(4) and a higher isotropic thermal parameter. On the contrary, each of the Ca(1), Ca(2), Ca(3), and Ca(5) is fully occupied by one Ca atom, and these positions are coordinated with seven, eight, eight, and six oxygen atoms, respectively.

Both powders used for sintering represent a hexagonal crystal system. Figure 1 presents the XRD patterns acquired for COM. HA and LAB. HA nanopowders. The diffraction patterns show a big similarity between both samples.

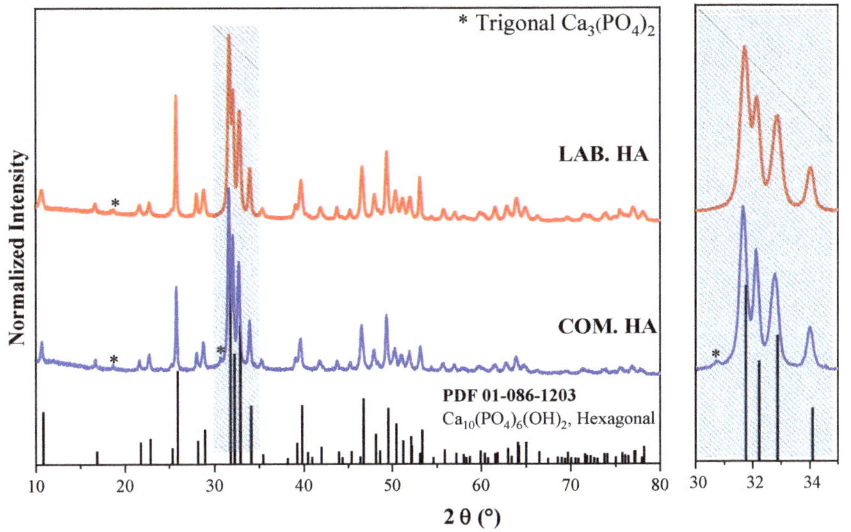

Figure 1. Powder XRD patterns of COM. HA and LAB. HA nano-crystalline calcium phosphates together with simulated PDF #01-086-1203 of hexagonal hydroxyapatite.

Reflexes within the 10–80 2θ range were identified and indexed to the hexagonal phase of $Ca10(PO_4)_6(OH)_2$ with s. g., *P63/m*. The wider reflexes, indicating smaller grains, are observed for the sample prepared in our laboratory. In turn, the diffraction patterns for the COM. HA display an additional line of very weak intensity at 30.5 2θ, corresponding to the trigonal phase of $Ca_3(PO_4)_2$. The studies show that the powder obtained by us is characterized by greater quality.

3.1.2. Morphology and Particle Size of Nanopowders by TEM Analysis

Using high-resolution TEM imaging, it was possible to determine the particle size of the investigated materials. As depicted in Figure 2, the micrographs of COM. HA and LAB. HA nanopowders exhibit big differences in the homogeneity, shape, and size of the nanopowder.

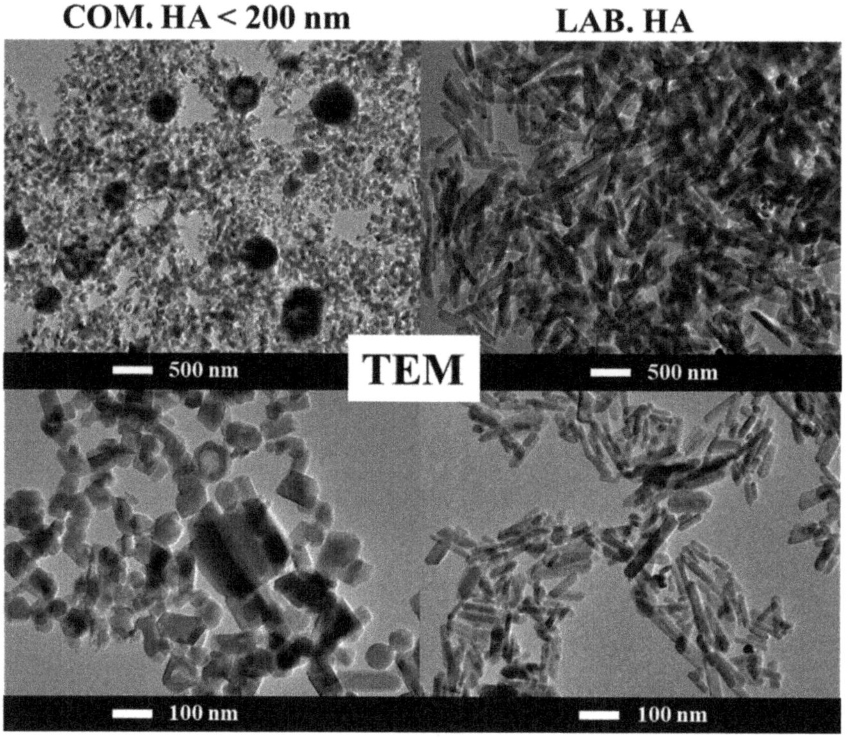

Figure 2. TEM image of COM. HA and LAB. HA nano-crystalline calcium phosphates.

The COM. HA possesses crystallites of different shapes, including spheres, rectangles, and ovals within the size range of 20–80 nm. The image obtained at low magnification shows that the material is quite homogeneous, the crystallites are well-separated from each other, and grain boundaries are clearly visible. However, spontaneously emerging agglomerates in the form of balls with a diameter of 200–500 nm are also visible. The micrographs shown in Figure S1 reveal the morphology of this nanopowder in detail. To check whether these agglomerates were not accidental, other commercially available powders were tested. The obtained results were identical. The elementary analysis by EDS (not presented here) was performed; however, it did not detect changes in the chemical composition of the big grains and the smaller ones. It is difficult to properly conclude why these large agglomerates are present in these commercially available powders.

In contrast, the LAB. HA demonstrates a significantly more homogeneous morphology without agglomerates in the form of balls. Nonetheless, the shape of the obtained crystallites is different; they form uniform rods with sizes ranging from 20 to 100 nm, as presented in detail in Figure S2.

The particle size distribution was analyzed more carefully. As can be seen in the histograms of both types of powder (Figure 3), the COM. HA powder in the majority is composed of particles with an average size of approximately 40 nm, while the sizes of the aggregates range from 175, 225, and 275 nm (marked with red arrows). For the nanopowders of LAB. HA, the particles in the form of rods reach an average size equal to 40–55 nm (length) and 19 nm (width). There is no doubt that LAB. HA nanopowder is more homogeneous.

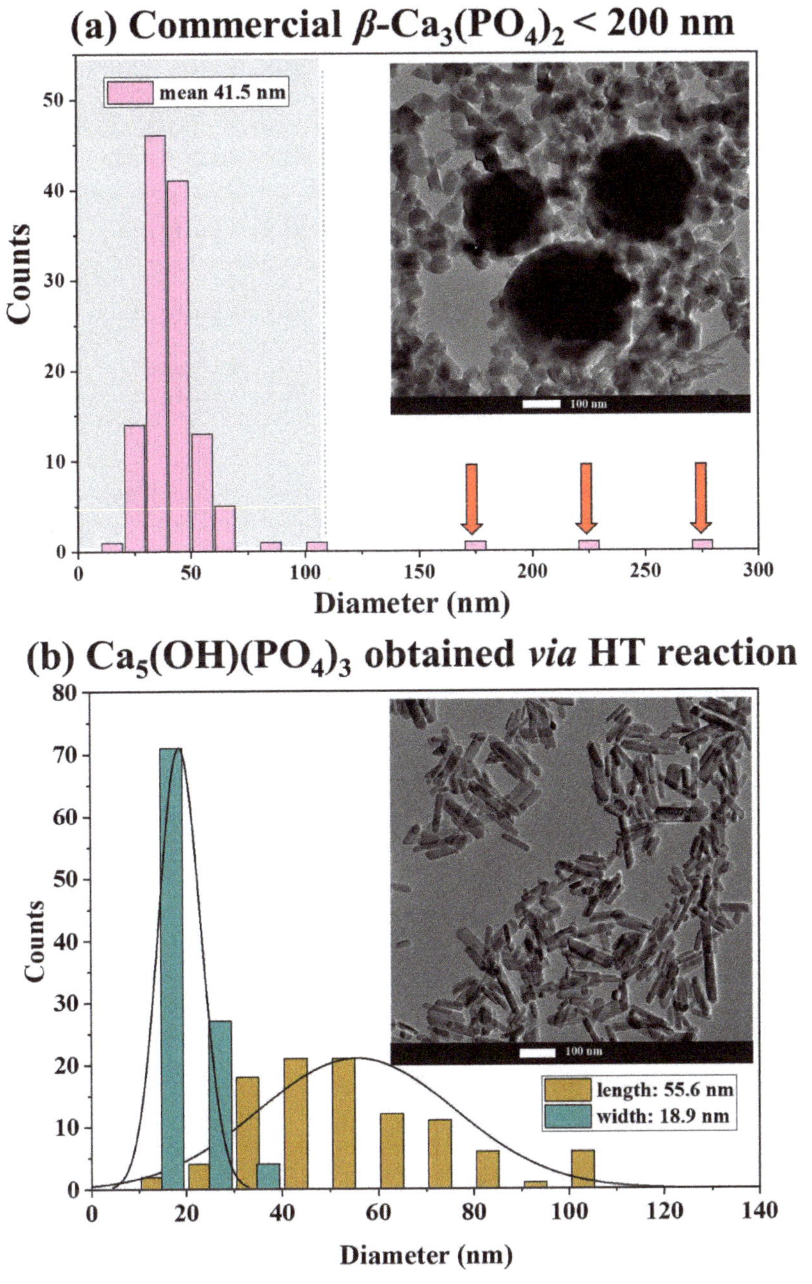

Figure 3. TEM images and particle size distribution of COM. HA (**a**) and LAB. HA (**b**).

3.2. Characterizations of Micro-Ceramics Fabricated by SPS

The current literature includes a few papers devoted to the fabrication of translucent and transparent polycrystalline ceramics based on tricalcium phosphate (β-TCP), hydroxyapatite (HA), and fluorapatite (FAP). Mainly, they are obtained by the SPS protocol, and the procedures exhibit variations in the applied sintering temperature, pressure, and dwell time. Furthermore, different synthesis methods were employed to obtain the initial materials, resulting in sintered bodies differing in quality.

Here, the influence on the quality of sintered materials of changing parameters, such as: (a) application of the sintering additive LiF; (b) temperature used during SPS; and (c) starting materials, i.e., commercially available nanopowder synthesized by us *via* the HT method, were investigated and are described below.

3.2.1. Influence of LiF Doping

These experiments were performed using only COM. HA nanopowder. Different amounts of LiF (0.25–1 wt.%) were added to the initial powder and ground in an agate mortar using a small amount of ethanol. The sintering conditions used in the four tests were identical in terms of pressure, moment of pressure application, temperature rise, and dwell time.

The SPS experiments were conducted based on the protocol proposed by Kato et al. [24]. The effect of the addition of LiF on SPS-produced calcium phosphate is illustrated in Figure 4. The best optical quality was observed when 0.25 and 0.5% of LiF were used.

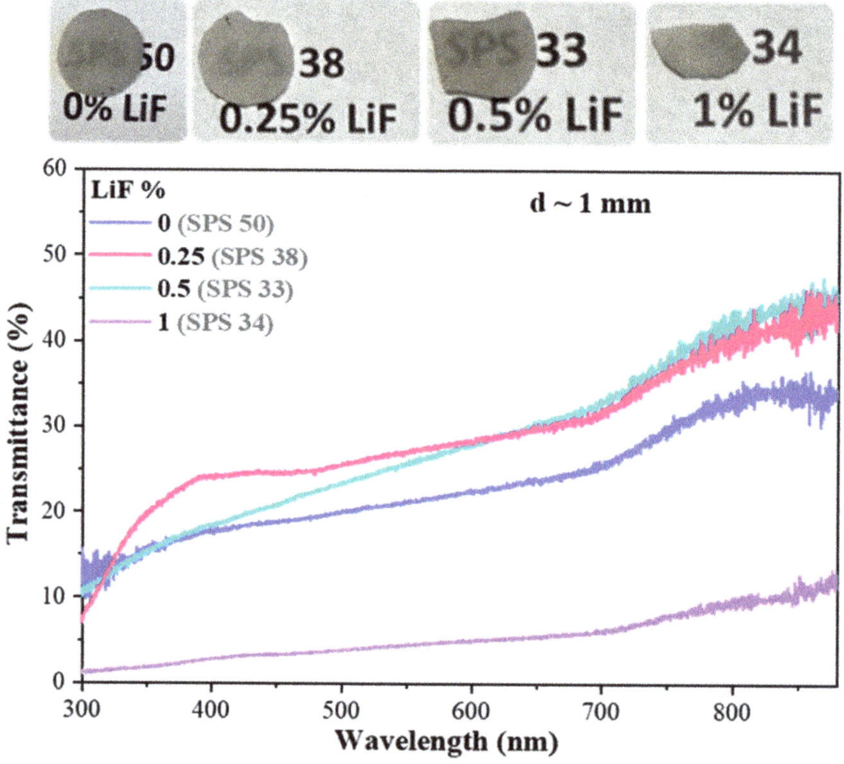

Figure 4. Photographs and total transmission spectra of translucent ceramics obtained by SPS at 1050 °C from COM. HA powder with different LiF contents. Transmittance normalized to ceramics' thickness = 1 mm.

The transmission spectra are similar, particularly within the infrared range of the spectrum, demonstrating a total transmittance of approximately 30% at 650 nm and 40–45% at 880 nm. Undoubtedly, the addition of LiF improves the quality and transmission of sintered bodies. However, when the concentration of the additive reaches 1 wt.%, the sample becomes completely opaque. This can be seen both in the photo and in the graph presenting the transmission spectra. Sporadically occurring black spots are mainly caused by the penetration of carbon into the ceramics from the graphite die during the sintering process.

Figure 5 presents the images of the surfaces of ceramics obtained under the conditions described above. The most homogeneous morphology is observed for samples with 0.25% and 0.5% of LiF, which is consistent with the observed light transmission. Both samples possess well-compacted grains of spherical shapes, and the grain boundaries between the microcrystals are clearly seen. The average grain size in a sample with 0.25% LiF is equal to 2 μm, and possible pores appear very rarely. In the case of ceramics with 0.5% LiF, the average grain size is equal to 3 μm, but sometimes there are places where the grains are smaller, with a size of approximately 1 μm. The pores between grains are also more common. For the sample without sintering additive, the grains are the smallest, with an average size of 1 μm. In turn, the lack of transparency in the sample with 1% LiF is reflected in the least favorable morphology. The lost transparency of this sample can be attributed to the growth of large grains reaching even 8.5 μm and the lack of homogeneity.

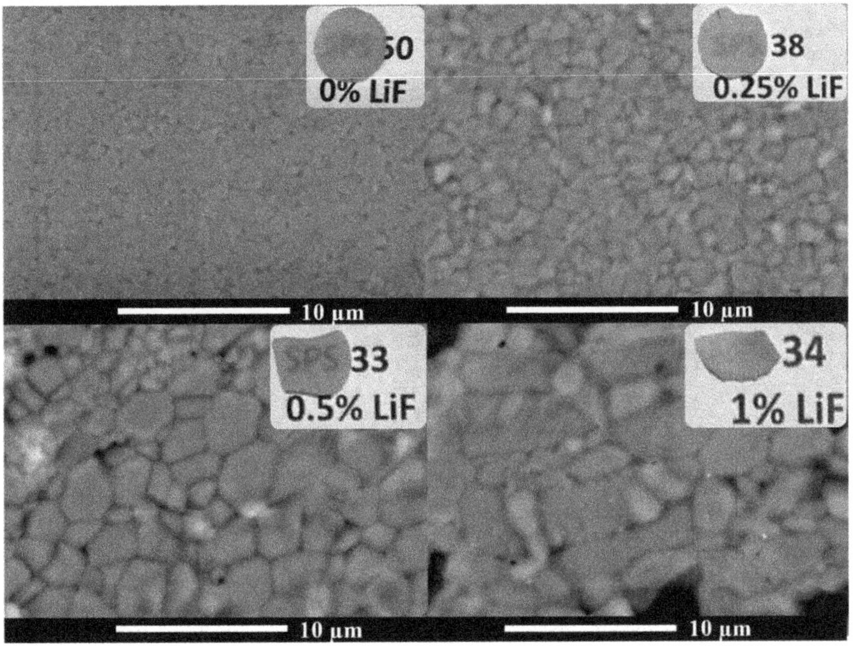

Figure 5. SEM micrographs of surfaces of ceramics obtained by SPS at 1050 °C from COM. HA nanopowder with different LiF contents.

The interesting and difficult-to-understand phenomenon, also observed but not discussed by Kato et al. [24], is the presence of two phases in all ceramics obtained as the result of the SPS process. Powder XRD patterns of all discussed here obtained by SPS at 1050 °C from COM. HA nanopowder with different LiF content are presented in Figure S3. Detailed information on the phase composition was gathered using Rietveld calculations. The ratio of HA to β-TCP varies depending on the percentage of LiF used. A rise in the hexagonal phase was observed with increasing LiF, and when 1% LiF was used, the amount of the hexagonal phase was close to 100%. However, in the case of the highest amount of hexagonal phase, the sintered body is completely opaque.

The values of relative densities range from 98.8 to 99.5%. The values for the samples containing 0.25% and 0.5% LiF are the lowest (98.8%). However, these two samples out of the four analyzed show the best light transmission. In the case of a sample with 1% LiF (density 99.5%), a compact microstructure is observed, but the bigger grain size makes it impossible to obtain transparency. Table 1 summarizes all the results discussed above.

Table 1. Based on Rietveld calculations, phase compositions of ceramics sintered by SPS at 1050 °C from COM. HA nanopowder with different LiF content; grain size observed by SEM; values of total transmittance monitored at 650 and 880 nm (transmittance normalized to ceramics' thickness = 1 mm); and measured density.

LiF wt. %	Phase 1—Main $Ca_5(PO_4)_3(OH)$ Hexagonal *P63/m* (No 176) HA	Phase 2 $Ca_3(PO_4)_2$ Trigonal *R3c* (No 161) *β*-TCP	Grain Size (μm)	Transmission in % at 650 nm	Transmission in % at 880 nm	Relative Density %
0	71.4%	28.6%	**0.41–2.1**	23.4	33.8	99.0
0.25	**77.7%**	**22.3%**	**0.55–3.2**	**29.7**	**42.8**	**98.8**
0.5	**81.8%**	**18.2%**	**2.1–5.4**	**30.5**	**45.6**	**98.8**
1	94.5%	5.5%	**1.8–8.5**	5.5	12.8	99.5

3.2.2. Influence of Temperature of SPS

The next stage of the research was to check the effect of the temperature used during the SPS process. Based on the results described above, we decided to use 0.25 wt.% LiF as a sintering aid. All the tests were performed using COM. HA nanopowder. In the literature, one can find a few papers on ceramics made from calcium phosphates and their procedures using different sintering temperatures (remark to the correcting person: temperature is only valid in the singular) [22,24,25,34]. However, no additives were used in these studies. All experiments were carried out using the same sintering parameters; only the temperature was modified.

Figure 6 presents the photographs and total transmission spectra of translucent ceramic samples obtained by SPS at different temperatures from COM. HA powder is mixed with 0.25 wt.% of LiF as a sintering additive. The best optical quality was observed for the samples sintered at 1000 °C. Although the quality of the sample obtained at 1050 °C is also good, the worst results were obtained for the ceramics sintered at the lowest (950 °C) and highest (1150 °C) temperatures. Both samples are almost black due to the effect of contamination by carbon originating from the graphite die.

The specimen obtained at 1100 °C is also darker than the two best samples. Temperature is a key parameter that plays a huge role in the process of manufacturing transparent ceramics. The analysis of the morphology of ceramics (Figure 7) and their optical quality shows a strong correlation. The best and most homogenous morphology is observed for the samples obtained at 1000 °C and 1050 °C. The grains of spherical shapes are well-compacted, and pores are observed to be extremely rare. In both cases, the average grain's size is equal to 2 μm. The specimen obtained at 950 °C creates crystallites of the smallest sizes (~1 μm), but the morphology is not very homogenous. An increase in the temperature of the SPS process resulted in a proportional increase in average grain size to even 6 or 7 μm at the highest temperature, i.e., 1150 °C.

Powder XRD patterns of all ceramics obtained by SPS at different temperatures from COM.HA nanopowder with 0.25% LiF are presented in Figure S4. The analysis revealed, as previously, the presence of HA and *β*-TCP, while the starting powder initially exhibited an almost pure phase of HA. The ratio of these two phases stays more or less on the same level around 3 (HA): 1 (*β*-TCP), regardless of temperature (Table 2). Its changes are not as significant as they were observed when the amount of LiF varied. A slight increase in the content of the HA phase was notable up to a temperature of 1000 °C, and then the amount of this phase decreased slightly.

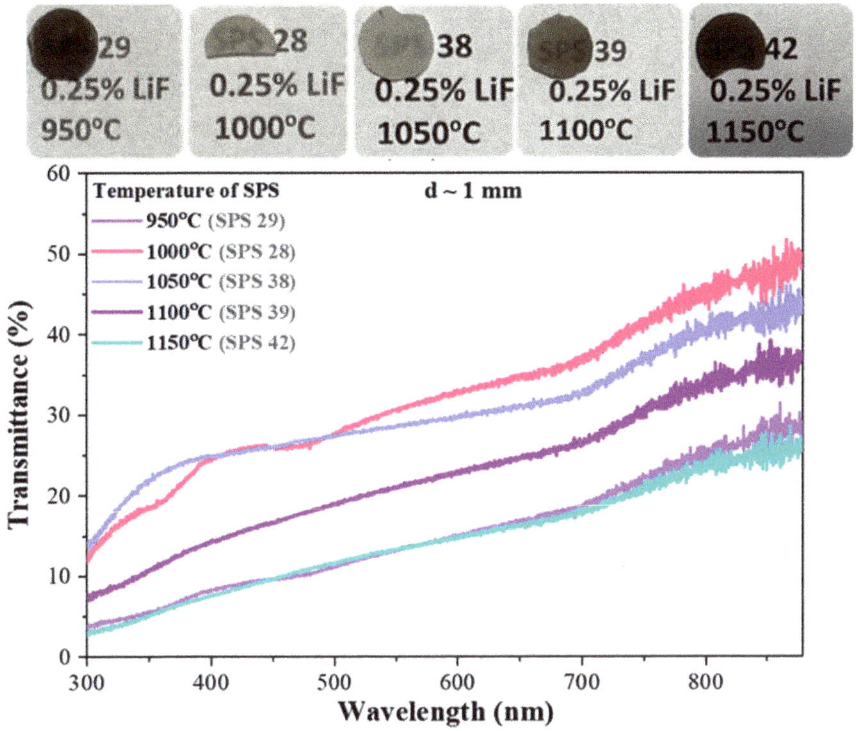

Figure 6. Photographs and total transmission spectra of ceramics obtained by SPS at different temperatures from COM. HA nanopowder with 0.25% LiF. Transmittance normalized to ceramics' thickness = 1 mm.

Table 2. Based on Rietveld calculations, phase compositions of ceramics sintered at different temperatures from COM.HA powder with 0.25% LiF; grain size observed by SEM; values of total transmittance monitored at 650 and 880 nm (transmittance normalized to ceramics thickness = 1 mm); and measured density.

Temperature of Sintering x °C/15 min	Phase 1—Main $Ca_5(PO_4)_3(OH)$ HA Hexagonal	Phase 2 $Ca_3(PO_4)_2$ β-TCP Trigonal	Grain Size (µm)	Transmission in % at 650 nm	Transmission in % at 880 nm	Relative Density %
950 °C	77.4%	22.6%	0.6–1.5	16.8	29.2	99.4
1000 °C	**79.3%**	**20.7%**	**0.7–2.6**	**34.8**	**48.7**	**99.3**
1050 °C	**77.7%**	**22.3%**	**0.55–3.2**	**29.7**	**42.8**	**98.8**
1100 °C	71.5%	28.5%	0.7–6.4	24.5	37.2	98.9
1150 °C	73.9%	26.1%	2.1–7.2	16.2	25.6	99.2

The values of measured densities range from 98.8 to 99.4%. The samples characterized by the best transmission of light and the best morphology show the highest density, as presented in Table 2.

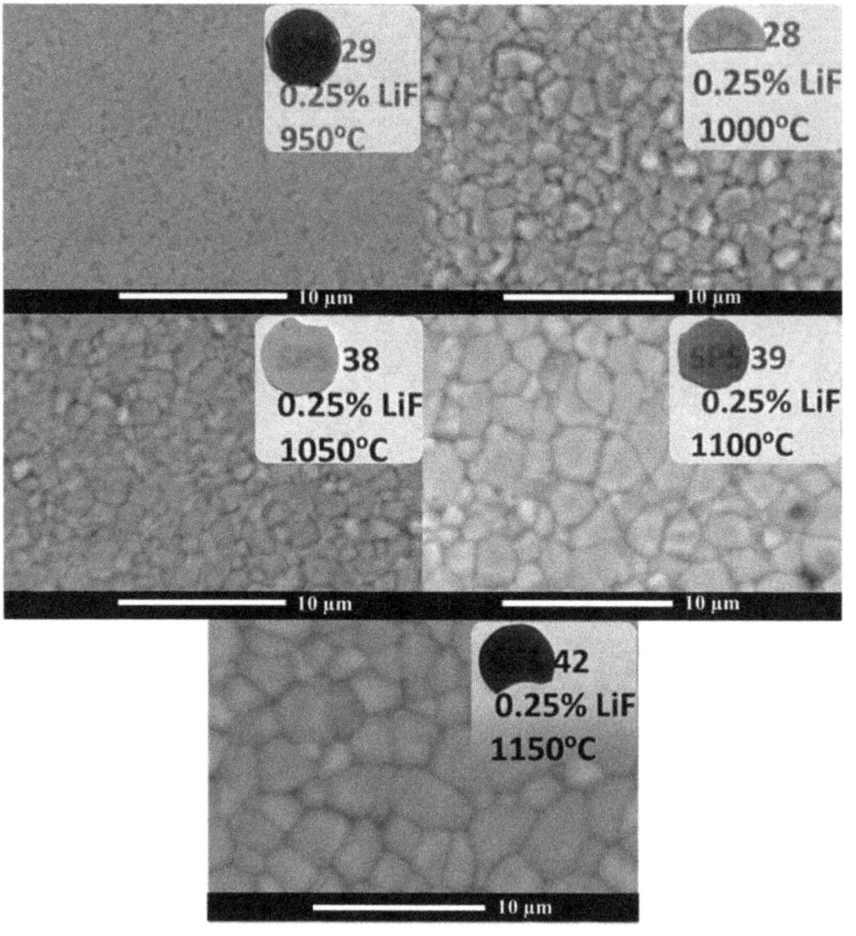

Figure 7. SEM micrographs of surfaces of ceramics obtained by SPS at different temperatures from COM.HA nanopowder with 0.25% LiF.

3.2.3. Influence of the Starting Material

According to the results presented above, the obtained sintered bodies did not exhibit the expected high level of transparency, similar to that which we could see in the reported papers [22,24]. The quality of the sintered specimens seems to be strongly influenced by the quality of the initial powder. All the experiments carried out by us, are based on the only calcium phosphate nanopowder available on our market.

So, taking into account our results indicating the optimal temperature and LiF content applied for the fabrication process, in the last step we decided to check how the starting materials influence the transparency of sintered materials. We prepared four samples from the nanopowdered hydroxyapatite synthesized in our laboratory by the hydrothermal method.

Two samples did not contain LiF, and the temperature of sintering was 1000 °C and 1050 °C, respectively. The next two were prepared at the same temperature, but using 0.25% LiF as a sintering aid.

Figure 8 presents photographs and total transmission spectra of translucent ceramic samples obtained by SPS from COM. HA and LAB. HA nanopowders. Without a doubt, it can be concluded that ceramics prepared from laboratory-made nanopowders are of much better quality and exhibit higher transparency. The sample from COM.HA powder without

LiF additive shows the worst light transmission, while the best specimens were obtained from LAB. HA with the addition of 0.25 wt.% LiF.

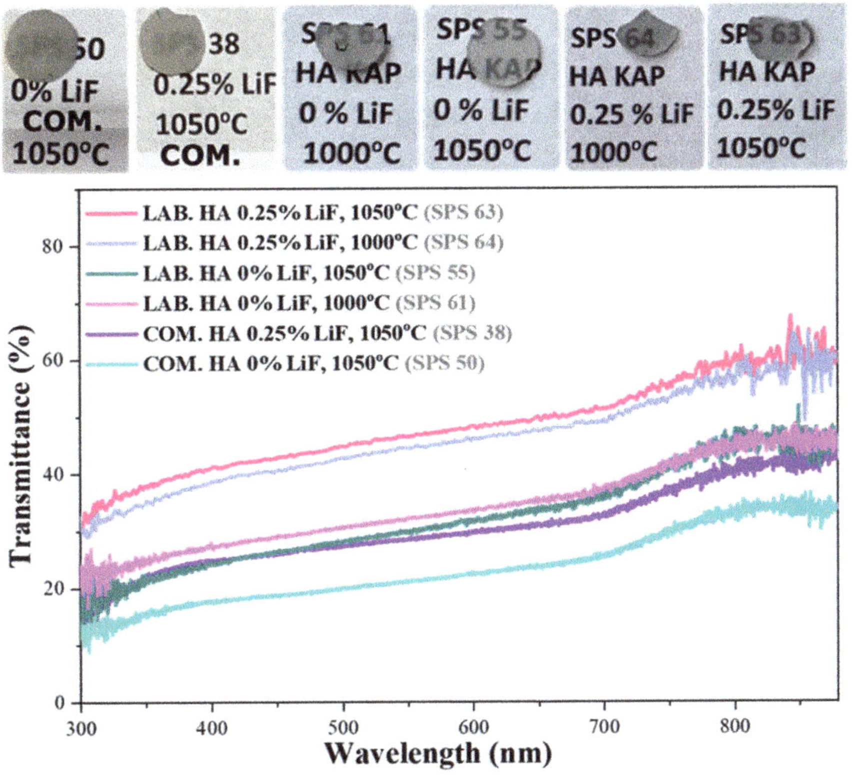

Figure 8. Photographs and total transmission spectra of ceramics obtained by SPS from COM. HA and LAB. HA nanopowders with and without the addition of LiF and sintered at 1000 °C or 1050 °C. Transmittance normalized to ceramic thickness = 1 mm.

Presented in Figure 9, SEM micrographs of surfaces of sintered bodies obtained from two different starting powders with and without the addition of LiF and sintered at 1000 °C or 1050 °C revealed that the temperature of 1000 °C applied for the sintering is too low or the time (15 min) is too short to obtain good homogeneity and uniform grain sizes. The sintered material consists of larger and smaller grains.

Powder XRD patterns of all ceramics obtained by SPS from COM. HA and LAB. HA nanopowders with and without the addition of LiF and sintered at 1000 °C or 1050 °C are presented in Figure S5. As previously stated, Rietveld's calculations (Table 3) made it possible to evaluate the ratio between HA and β-TCP phases present in all obtained ceramics. For the LAB. HA, the amount of hexagonal phase is higher than that calculated for specimens obtained from COM. HA powder, especially for both ceramics fabricated using 0.25 wt.% LiF, for which the values reach almost 90%. For these two samples, the highest values of transmission of light, almost 60%, were obtained.

In the case of the samples from two types of starting materials, the values of all measured densities are above 99%. The highest density was observed for the ceramics from LAB. HA with 0.25% LiF sintered at 1050 °C (Table 3), which corresponds to the best results obtained from SEM and transmission analysis (Figure S6).

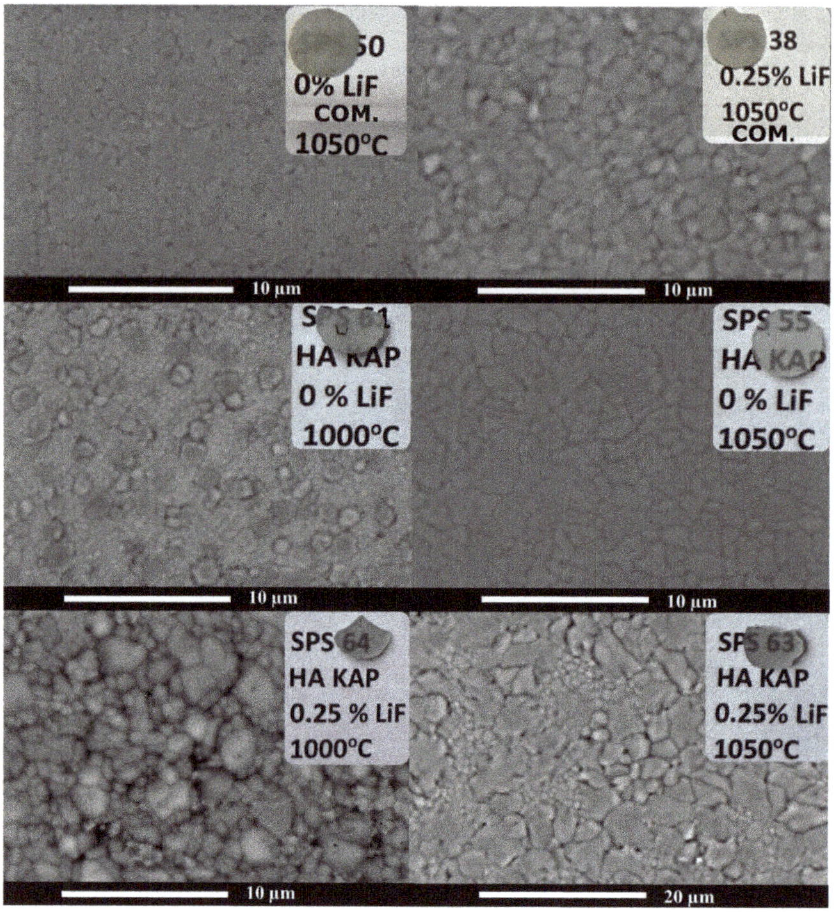

Figure 9. SEM micrographs of surfaces of ceramics obtained by SPS from COM. HA and LAB. HA nanopowders with and without the addition of LiF and sintered at 1000 °C or 1050 °C.

Table 3. Based on Rietveld calculations, phase compositions of ceramics obtained by SPS from COM. HA and LAB. HA nanopowders with and without application of LiF and sintered at 1000 °C or 1050 °C; grain size from SEM; values of total transmittance monitored at 650 and 880 nm (transmittance normalized to ceramics' thickness = 1 mm); and measured density.

Starting Material	LiF wt. %	Temp. of SPS x °C/ 15 min	Phase 1 HA Hexagonal	Phase 2 β-TCP Trigonal	Grain Size (µm)	Transmission in % at 650 nm	Transmission in % at 880 nm	Relative Density %
COM. HA	0	1050 °C	70.3%	29.7%	0.48–2.1	24.1	34.4	99.0
	0.25		77.7%	22.3%	0.55–3.2	29.7	42.8	98.8
LAB. HA	0	1000 °C	87.2%	12.8%	0.29–0.35 0.9–1.5	35.3	45.9	100.2
		1050 °C	84.6%	15.4%	0.7–2.7	33.3	45.1	99.6
	0.25	**1000 °C**	**89.8%**	**10.2%**	**0.4–1** **2.2–3.2**	**47.3%**	**59.1%**	**99.8**
		1050 °C	**89.4%**	**10.6%**	**0.5–0.8** **2.4–5.0**	**50%**	**59.3%**	**99.9**

4. Conclusions

Two types of nanopowders, i.e., commercially available (COM. HA) and laboratory-made by us (LAB. HA) *via* the hydrothermal (HT) protocol, were used to fabricate non-cubic hydroxyapatite $Ca_{10}(PO_4)_6(OH)_2$-based microcrystalline ceramics. All sintering procedures were performed *via* the SPS process, which is an efficient method of consolidating polycrystalline materials at relatively low temperatures in a short time. The effect of affected parameters such as the addition of LiF sintering agent, the temperature applied during the SPS process (in the range 950–1100 °C), and the quality of the starting nanopowders, on the optical quality of sintered ceramics were checked. XRD, microscopic (TEM and SEM) techniques, and light transmission measurements were used to characterize both the starting materials as well as the quality of the sintered samples. Highly transparent HA micro-ceramics were obtained from nano-crystalline powder obtained *via* the hydrothermal method. The addition of LiF as a sintering aid has a positive effect on the quality of ceramics when added in an amount of 0.25–0.5 wt.%, while the most favorable temperature for SPS sintering of this material was 1000–1050 °C for 15 min under 100 MPa pressure. These optimal conditions result in well-compacted dense micro-ceramics with a grain size of ~2–3 μm and transmission at 880 nm of ~60%. The research highlighted the crucial role of the quality of the starting materials on the quality and optical transparency of polycrystalline ceramics sintered by the SPS method. The values of measured densities stay in agreement with the results obtained from SEM analysis and measurements of light transmission.

Subsequent investigations toward enhancing transparency through the exploration of variations in applied pressure as well as the fabrication of HA ceramics doped by RE^{3+} ions are the subject of our present research.

Supplementary Materials: The following supporting information can be downloaded at: https://www.mdpi.com/article/10.3390/ceramics7020040/s1, Figure S1: TEM micrographs of COM. HA, nano-crystalline calcium phosphates; Figure S2: TEM micrographs of LAB. HA, nano-crystalline calcium phosphates; Figure S3: Powder XRD patterns of ceramics obtained by SPS at 1050 °C from COM. HA nano-powder with different LiF content as well as simulated patterns of hexagonal and trigonal phases of HA; Figure S4: Powder XRD patterns of ceramics obtained by SPS at different temperature from COM.HA nano-powder with 0.25% of LiF as well as simulated patterns of hexagonal and trigonal phases of HA; Figure S5: Powder XRD patterns of ceramics obtained by SPS from COM. HA and LAB. HA nano-powders with and without the addition of LiF and sintered at 1000 °C or 1050 °C as well as simulated patterns of hexagonal and trigonal phases of HA; Figure S6: SEM micrographs of surfaces of ceramics obtained at 1050 °C from COM.HA nano-powder with 0.25% of LiF, which shows the best transparency.

Author Contributions: Conceptualization, K.A.P., S.C., V.G., G.F., Y.G., G.B. and M.G.; Data curation, K.A.P., S.C. and M.G.; Formal analysis, K.A.P., S.C., V.G., G.F., Y.G., G.B. and M.G.; Funding acquisition, K.A.P. and M.G.; Investigation, K.A.P., S.C., V.G., G.F., Y.G., G.B. and M.G.; Methodology, S.C., V.G. and G.F.; Project administration, M.G.; Resources, S.C., V.G. and G.F.; Supervision, G.F., G.B. and M.G.; Validation, G.F. and G.B.; Visualization, K.A.P. and M.G.; Writing—original draft, K.A.P. and M.G.; Writing—review and editing, S.C., V.G., G.F., Y.G. and G.B. All authors have read and agreed to the published version of the manuscript.

Funding: The research was co-financed by the ARQUS (European University Alliance) with support from UCBLyon1. M.G. would like to thank the Polish National Agency for Academic Exchange (NAWA) for M. Bekker's scholarship No. PPN/BEK/2020/1/00350/U/00001. K.A.P. would like to thank the Ministry of Education and Science in Poland for Grant No. DWD/5/0361/2021 in the framework of the Implementation Doctorate Program.

Institutional Review Board Statement: Not applicable.

Informed Consent Statement: Not applicable.

Data Availability Statement: The data will be made available upon request.

Conflicts of Interest: The authors declare no conflicts of interest.

References

1. Ikesue, A.; Aung, L.A. (Eds.) *Processing of Ceramics: Breakthroughs in Optical Materials*; Wiley: Hoboken, NJ, USA, 2021.
2. Goldstein, A.; Krell, A. Transparent ceramics at 50: Progress made and further prospects. *J. Am. Ceram. Soc.* **2016**, *99*, 3173–3197. [CrossRef]
3. Goldstein, A.; Andreas Krell, A.; Burshtein, Z. *Transparent Ceramics: Materials, Engineering, and Applications*; Wiley: Hoboken, NJ, USA, 2020; ISBN 978-1-119-42949-4.
4. Krell, A.; Hutzler, T.; Klimke, J. Transmission physics and consequences for materials selection, manufacturing, and applications. *J. Eur. Ceram. Soc.* **2009**, *29*, 207–222. [CrossRef]
5. Hostaša, J.; Picelli, F.; Hříbalová, S.; Nečina, V. Sintering aids, their role and behaviour in the production of transparent ceramics. *Open Ceram.* **2021**, *7*, 100137. [CrossRef]
6. Frage, N.; Cohen, S.; Meir, S.; Kalabukhov, S.; Dariel, M.P. Spark plasma sintering (SPS) of transparent magnesium-aluminate spinel. *J. Mater. Sci.* **2007**, *42*, 3273–3275. [CrossRef]
7. Nečina, V.; Pabst, W. Comparison of the effect of different alkali halides on the preparation of transparent $MgAl_2O_4$ spinel ceramics via spark plasma sintering (SPS). *J. Eur. Ceram. Soc.* **2020**, *40*, 6043–6052. [CrossRef]
8. Ikesue, A.; Kinoshita, T.; Kamata, K.; Yoshida, K. Fabrication and optical properties of high-performance polycrystalline Nd: YAG ceramics for solid-state lasers. *J. Am. Ceram. Soc.* **1995**, *78*, 1033–1040. [CrossRef]
9. Yagi, H.; Yanagitani, Y.; Takaichi, K.; Ueda, K.; Kaminskii, A. Characterizations and laser performances of highly transparent Nd^{3+}:$Y_3Al_5O_{12}$ laser ceramics. *Opt. Mater.* **2007**, *29*, 1258–1262. [CrossRef]
10. Talimian, A.; Pouchly, V.; El-Maghraby, H.F.; Maca, K.; Galusek, D. Transparent magnesium aluminate spinel: Effect of critical temperature in two-stage spark plasma sintering. *J. Eur. Ceram. Soc.* **2020**, *40*, 2417–2425. [CrossRef]
11. Grasso, S.; Yoshida, H.; Porwal, H.; Sakka, Y.; Reece, M. Highly transparent α-alumina obtained by low cost high pressure SPS. *Ceram. Int.* **2013**, *39*, 3243–3248. [CrossRef]
12. Kim, B.N.; Hiraga, K.; Morita, K.; Yoshida, H. Spark plasma sintering of transparent alumina. *Scr. Mater.* **2007**, *57*, 607–610. [CrossRef]
13. Sarthou, J.; Aballea, P.; Patriarche, G.; Serier-Brault, H.; Suganuma, A.; Gredin, P.; Mortier, M.; Riman, R. Wet-route synthesis and characterization of Yb:CaF_2 optical ceramics. *J. Am. Ceram. Soc.* **2016**, *99*, 1992–2000. [CrossRef]
14. Ivanov, M.; Kalinina, E.; Kopylov, Y.; Kravchenko, V.; Krutikova, I.; Kynast, U.; Li, J.; Leznina, M.; Medvedev, A. Highly transparent Yb-doped $(La_xY_{1-x})_2O_3$ ceramics prepared through colloidal methods of nanoparticles compaction. *J. Eur. Ceram. Soc.* **2016**, *36*, 4251–4259. [CrossRef]
15. Ikesue, A.; Aung, L.A.; Lupei, V. *Ceramic Lasers*; Cambridge University Press: Cambridge, UK, 2013.
16. Sobota, P.; Guzik, M.; Garnier, V.; Fantozzi, G.; Sobota, M.; Tomaszewicz, E.; Guyot, Y.; Boulon, G. Fabrication of Y_6MoO_{12} molybdate ceramics: From synthesis of cubic nanopowder to sintering. *Ceram. Int.* **2020**, *46*, 4619–4633. [CrossRef]
17. Wilk (Bieza), M.; Tomaszewicz, E.; Siczek, M.; Guyot, Y.; Boulon, G.; Guzik, M. The first characterization of cubic Nd^{3+}-doped mixed La_2MoWO_9 in micro-crystalline powders and translucent micro-ceramics. *J. Mater. Chem. C* **2022**, *10*, 10083–10098. [CrossRef]
18. Bieza, M.; Guzik, M.; Tomaszewicz, E.; Guyot, Y.; Lebbou, K.; Zych, E.; Boulon, G. Toward optical ceramics based on cubic Yb^{3+} rare earth ion-doped mixed molybdato- tungstates: Part I—Structural characterization. *J. Phys. Chem. C* **2017**, *121*, 13290–13302. [CrossRef]
19. Bieza, M.; Guzik, M.; Tomaszewicz, E.; Guyot, Y.; Lebbou, K.; Boulon, G. Toward optical ceramics based on cubic Yb^{3+} rare earth ion-doped mixed molybdate-tungstates: Part II—Spectroscopic characterization. *J. Phys. Chem. C* **2017**, *121*, 13303–13313. [CrossRef]
20. Prokop, K.A.; Siczek, M.; Tomaszewicz, E.; Rola, K.; Guyot, Y.; Boulon, G.; Guzik, M. Structural ordering studies of Nd^{3+} ion in cubic $M_3Y(PO_4)_3$ (M = Sr^{2+} or Ba^{2+}) perovskites. First translucent ceramics from micro-crystalline cubic powders. *Ceram. Int.* **2024**, *50*, 8042–8056. [CrossRef]
21. Kawagoe, D.; Ioku, K.; Fujimori, H.; Goto, S. Transparent β-tricalcium phosphate ceramics prepared by spark plasma sintering. *J. Ceram. Soc. Jpn.* **2004**, *112*, 462–463. [CrossRef]
22. Kim, B.N.; Horiuchi, N.; Dash, A.; Kim, Y.W.; Morita, K.; Yoshida, H.; Ji-Guang, L.; Sakka, Y. Spark plasma sintering of highly transparent hydroxyapatite ceramics. *J. Jpn. Soc. Powder Powder Metall.* **2017**, *64*, 547–551. [CrossRef]
23. Furuse, H.; Kato, D.; Morita, K.; Suzuki, T.S.; Kim, B.N. Characterization of Transparent Fluorapatite Ceramics Fabricated by Spark Plasma Sintering. *Materials* **2022**, *15*, 8157. [CrossRef]
24. Kato, T.; Nakauchi, D.; Kawaguchi, N.; Yanagida, T. Optical and X-ray-induced luminescence properties of calcium phosphate of transparent ceramic form. *Sens. Mater.* **2020**, *32*, 1411–1417. [CrossRef]
25. Ioku, K.; Kamitakahara, M. Hydroxyapatite ceramics for medical application prepared by hydrothermal method. *Phosphorus Res. Bull.* **2009**, *23*, 25–30. [CrossRef]
26. Kotobuki, N.; Kawagoe, D.; Nomura, D.; Katou, Y.; Muraki, K.; Jujimori, H.; Goto, S.; Ioku, K.; Ohgushi, H. Observation and quantitative analysis of rat bone marrow stromal cells cultured in vitro on newly formed transparent β-tricalcium phosphate. *J. Mater. Sci. Mater Med.* **2006**, *17*, 33–41. [CrossRef]
27. Akiyama, J.; Sato, Y.; Taira, T. Laser ceramics with rare-earth-doped anisotropic materials. *Opt. Lett.* **2010**, *35*, 3598–3600. [CrossRef]

28. Akiyama, J.; Sato, Y.; Taira, T. Laser demonstration of diode-pumped Nd^{3+}-doped fluorapatite anisotropic ceramics. *Appl. Phys. Express.* **2011**, *4*, 022703-1–022703-3. [CrossRef]
29. Furuse, H.; Horiuchi, N.; Kim, B.N. Transparent non-cubic laser ceramics with fine microstructure. *Sci. Rep.* **2019**, *9*, 2–3. [CrossRef]
30. Furuse, H.; Okabe, T.; Shirato, H.; Kato, D.; Horiuchi, N.; Morita, K.; Kim, B.N. High-optical-quality non-cubic Yb^{3+}-doped $Ca_{10}(PO_4)_6F_2$ (Yb:FAP) laser ceramics. *Opt. Mater. Express* **2021**, *11*, 1756. [CrossRef]
31. Eriksson, M.; Liu, Y.; Hu, J.; Gao, L.; Nygren, M.; Shen, Z. Transparent hydroxyapatite ceramics with nanograin structure prepared by high-pressure spark plasma sintering at the minimized sintering temperature. *J. Eur. Ceram. Soc.* **2011**, *31*, 1533–1540. [CrossRef]
32. Rubat du Merac, M.; Kleebe, H.J.; Müller, M.M.; Reimanis, I.E. Fifty years of research and development coming to fruition; Unraveling the complex interactions during processing of transparent magnesium aluminate ($MgAl_2O_4$) spinel. *J. Am. Ceram. Soc.* **2013**, *96*, 3341–3365. [CrossRef]
33. Frage, N.; Kalabukhov, S.; Sverdlov, N.; Ezersky, V.; Dariel, M.P. Densification of transparent yttrium aluminum garnet (YAG) by SPS processing. *J. Eur. Ceram. Soc.* **2010**, *30*, 3331–3337. [CrossRef]
34. Hu, Z.; Xu, X.; Wang, J.; Liu, P.; Li, D.; Wang, X.; Zhang, J.; Xu, J.; Tang, D. Fabrication and spectral properties of Dy:Y_2O_3 transparent ceramics. *J. Eur. Ceram. Soc.* **2018**, *38*, 1981–1985. [CrossRef]
35. Hu, Z.; Xu, X.; Wang, J.; Liu, P.; Li, D.; Wang, X.; An, L.; Zhang, J.; Xu, J.; Tang, D. Spark plasma sintering of Sm^{3+} doped Y_2O_3 transparent ceramics for visible light lasers. *Ceram. Int.* **2017**, *43*, 12057–12060. [CrossRef]
36. Alombert-Goget, G.; Guyot, Y.; Guzik, M.; Boulon, G.; Ito, A.; Goto, T.; Yoshikawa, A.; Kikuchi, M. Nd^{3+}-doped Lu_2O_3 transparent sesquioxide ceramics elaborated by the Spark Plasma Sintering (SPS) method. Part 1: Structural, thermal conductivity and spectroscopic characterization. *Opt. Mater.* **2015**, *41*, 3–11. [CrossRef]
37. Jiang, N.; Xie, R.-J.; Liu, Q.; Li, J. Fabrication of sub-micrometer MgO transparent ceramics by spark plasma sintering. *J. Eur. Ceram. Soc.* **2017**, *37*, 4947–4953. [CrossRef]
38. Jokić, B.; Mitrić, M.; Radmilović, V.; Drmanić, S.; Petrović, R.; Janaćković, D. Synthesis and characterization of monetite and hydroxyapatite whiskers obtained by a hydrothermal method. *Ceram. Int.* **2011**, *37*, 167–173. [CrossRef]
39. Syukkalova, E.A.; Sadetskaya, A.V.; Demidova, N.D.; Bobrysheva, N.P.; Osmolowsky, M.G.; Voznesenskiy, M.A.; Osmolovskaya, O.M. The effect of reaction medium and hydrothermal synthesis conditions on morphological parameters and thermal behavior of calcium phosphate nanoparticles. *Ceram. Int.* **2021**, *47*, 2809–2821. [CrossRef]
40. Grigoraviciute-Puroniene, I.; Tsuru, K.; Garskaite, E.; Stankeviciute, Z.; Beganskiene, A.; Ishikawa, K.; Kareiva, A. A novel wet polymeric precipitation synthesis method for monophasic β-TCP. *Adv. Powder Technol.* **2017**, *28*, 2325–2331. [CrossRef]
41. Han, Y.; Li, S.; Wang, X.; Chen, X. Synthesis and sintering of nanocrystalline hydroxyapatite powders by citric acid sol-gel combustion method. *Mater. Res. Bull.* **2004**, *39*, 25–32. [CrossRef]
42. He, W.; Xie, Y.; Xing, Q.; Ni, P.; Han, Y.; Dai, H. Sol-gel synthesis of biocompatible Eu^{3+}/Gd^{3+} co-doped calcium phosphate nanocrystals for cell bioimaging. *J. Lumin.* **2017**, *192*, 902–909. [CrossRef]
43. Anee, T.K.; Ashok, M.; Palanichamy, M.; Kalkura, S.N. A novel technique to synthesize hydroxyapatite at low temperature. *Mater. Chem. Phys.* **2003**, *80*, 725–730. [CrossRef]
44. Han, J.K.; Song, H.Y.; Saito, F.; Lee, B.T. Synthesis of high purity nano-sized hydroxyapatite powder by microwave-hydrothermal method. *Mater. Chem. Phys.* **2006**, *99*, 235–239. [CrossRef]
45. Kalita, S.J.; Verma, S. Nanocrystalline hydroxyapatite bioceramic using microwave radiation: Synthesis and characterization. *Mater. Sci. Eng. C* **2010**, *30*, 295–303. [CrossRef] [PubMed]
46. Choi, D.; Kumta, P.N. Mechano-chemical synthesis and characterization of nanostructured β-TCP powder. *Mater. Sci. Eng. C* **2007**, *27*, 377–381. [CrossRef]
47. Bricha, M.; Belmamouni, Y.; Essassi, E.M.; Ferreira, J.M.F.; El Mabrouk, K. Surfactant-assisted hydrothermal synthesis of hydroxyapatite nanopowders. *J. Nanosci. Nanotechnol.* **2012**, *12*, 8042–8049. [CrossRef] [PubMed]
48. Bilton, M.; Milne, S.J.; Brown, A.P. Comparison of Hydrothermal and Sol-Gel Synthesis of Nano-Particulate Hydroxyapatite by Characterisation at the Bulk and Particle Level. *Open J. Inorg. Non Met. Mater.* **2012**, *2*, 1–10. [CrossRef]
49. Ma, G.; Liu, X.Y. Hydroxyapatite: Hexagonal or Monoclinic? *Cryst. Growth Des.* **2009**, *9*, 2991–2994. [CrossRef]
50. Dickens, B.; Schroeder, L.W.; Brown, W.E. Crystallographic studies of the role of Mg as a stabilizing impurity in β-$Ca_3(PO_4)_2$. The crystal structure of pure β-$Ca_3(PO_4)_2$. *J. Solid State Chem.* **1974**, *10*, 232–248. [CrossRef]

ceramics

MDPI

Article

Beyond Scanning Electron Microscopy: Comprehensive Pore Analysis in Transparent Ceramics Using Optical Microscopy

Francesco Picelli [1,2], Jan Hostaša [1], Andreana Piancastelli [1], Valentina Biasini [1], Cesare Melandri [1] and Laura Esposito [1,*]

[1] CNR Institute of Science, Technology, and for Sustainability Ceramics, Via Granarolo 64, 48018 Faenza, Italy; francesco.picelli@issmc.cnr.it (F.P.); jan.hostasa@issmc.cnr.it (J.H.); andreana.piancastelli@issmc.cnr.it (A.P.); valentina.biasini@issmc.cnr.it (V.B.); cesare.melandri@issmc.cnr.it (C.M.)

[2] Department of Chemistry, Life Sciences and Environmental Sustainability, Università degli Studi di Parma, Parco Area delle Scienze 17/A, 43124 Parma, Italy

* Correspondence: laura.esposito@issmc.cnr.it

Abstract: Developing an effective method of quantifying defects in the bulk of transparent ceramics is a challenging task that could facilitate their widespread use as a substitute for single crystals. Conventionally, SEM analysis is used to examine the microstructure but it is limited to the material surface. On the other hand, optical transmittance assesses material quality, but does not provide information on the size and concentration of defects. In this study, we illustrate the use of a digital optical microscope for the non-destructive, precise, and rapid analysis of residual porosity in transparent ceramics. YAG-based ceramics doped with Yb have been selected for this study because they are used as laser gain media, an application that requires virtually defect-free components. Different production processes were used to produce YAG samples, and the digital optical microscope analysis was used to compare them. This analysis was shown to be effective and precise to measure the size and concentration of the residual pores. In addition, the comparison of samples obtained with different production processes showed that the size and distribution of the residual porosity is affected by the drying step of the powders before shaping by pressing, as well as by the sintering aids used to ease the densification. It also showed that the transmittance is influenced by both the total volume and the concentration of the pores.

Keywords: transparent ceramics; YAG; optical microscope; residual porosity; pore quantification

1. Introduction

Transparent polycrystalline ceramics can replace single crystals in optical applications, such as scintillators, light converters, and, in the case of YAG-based compositions, solid-state laser gain media [1–3]. The ceramic process offers several advantages over the melt growth methods used to prepare single crystals: in particular, faster production times, increased doping levels and the opportunity to design profiles and structures before sintering. Although significant progress has been made in improving the optical quality, ceramics often scatter light more than single crystals due to microstructural features like pores, impurities, grain boundaries, and birefringence effects [4,5]. Ceramics are polycrystalline materials with a microstructure that forms during the sintering process and that is strongly influenced by the shaping process. A green compact is formed by particles assembled together through pressing, casting, or other shaping techniques and is then densified during the sintering process. Transparent ceramics need to reach full density during sintering; i.e., pores should be completely removed. To promote complete densification, during the shaping step, compact particle packing should be obtained and the porosity should be homogeneous in size and shape. In addition, to ease the densification, advanced sintering techniques (e.g., vacuum sintering, sintering in the atmosphere of a specific gas, or pressure-assisted sintering) are often used, separately or together, and are often coupled with suitable sintering aids [6–9]. The production of transparent ceramics is therefore a complex process that involves several processing steps, and each of them could be responsible for defect formation. All these features make it challenging to identify the source of any defects that may arise. Furthermore, most of the defects can be identified only at the end of the production process and after sample polishing.

To ensure competitiveness with a single crystal, it is important for the ceramic to be free from defects. The presence of secondary phases and residual porosity are among the most common defects that need to be addressed [10,11]. Yet, of all the factors that contribute to light scattering, the most significant cause of light attenuation in transparent ceramics is pore scattering [4,10,12]. This is due to the major difference between the refractive index of the ceramic matrix and the defects: the difference between the ceramic phase and the pores is about one order of magnitude higher compared to that of two oxide phases. The difference between the refractive indexes is determinative of the attenuation of the effects of scattering centres in a transparent medium.

Measuring the concentration and dimensional distribution of defects is a daily challenge in the study of transparent ceramics. Frequently used analytical techniques include scanning electron microscopy (SEM) and measurements of optical transmittance [13,14]. SEM analysis can be used to examine the microstructure and determine the nature of the defects. However, this analysis is limited to the surface of the material, and, in the case of high optical quality, the concentration of defects decreases to such an extent that many images would need to be taken for a thorough analysis. In addition, it is a destructive analysis because several layers of the material should be cut, polished, and observed. Optical transmittance measurement provides a rapid means of assessing the optical quality of the material. In this case, the entire thickness of the sample is analysed, providing more information about the material's performance under real-world conditions. However, transmittance alone does not provide accurate information about the size and distribution of the defects.

There is therefore a need for a non-destructive experimental technique for the quantitative evaluation of the concentration and size of defects present in a volume that is representative of the entire sample. Many groups have proposed methods to characterize residual porosity in transparent materials. In 1991, Lopez et al. used the custom setup of a 1.06 μm YAG laser and an IR camera to capture the pores inside a fluoride glass [15]. More recently, Stuer et al. [16] proposed a characterization technique utilizing 3D focused ion beam (FIB) tomography coupled with SEM to slice and reconstruct pore positions in transparent alumina. This technique is well suited to the characterization of nanometric-sized pores (<150 nm in diameter), resulting in a relatively small analysed volume of approximately 69 μm^3, due to the high magnification used. Another drawback of this technique is that it is destructive. Boulesteix et al. [17] used confocal laser scanning microscopy (CLSM) to reconstruct the pore distribution within Nd:YAG. However, due to the low resolution of CLSM, an additional SEM analysis over many images was required to determine the pore size. Kosyanov et al. and Yavetskiy et al. used CLSM to calculate both the pore size and their concentration in Nd:YAG [18,19]. These techniques are non-destructive and allow for the analysis of larger volumes, approximately ~1.25×10^5 μm^3 according to Yavetskiy et al. and 2×10^5 μm^3 according to Boulesteix et al. The average pore size detected for Yavetskiy was from 150 to 200 nm, and for Boulesteix, it was from 0.5 to 0.7 μm [17,19].

All these techniques, which involve the use of SEM or CLSM, are subject to the so-called "corpuscle problem", which was raised and solved by Wicksell in 1925 [20]. The "corpuscle problem" addresses the problem of determining the distribution of spherical particles from planar sections by reconstructing the true size of pseudo-spherical features on a flat image that represent a cross-section of the material under examination. When arbitrarily slicing a transparent ceramic, in the search for defects, the chances of having all the defects sliced in their centre is zero. It has also been observed that large defects are statistically more prone to be sliced compared to smaller ones. As a result, the observed diameter of the defects is generally smaller than the real diameter of the defects and the smaller portion of defects can be under-assessed. It is worth noting that this issue is not limited to defect quantification in transparent polycrystalline ceramics, but it is also present in various other branches of natural sciences, ranging from biology to astronomy. Other researchers have also contributed to the refinement of Wicksell's results [20], including Saltykov [21], Cruz-Orive [22], and, more recently, Pabst et al. [23].

Jin et al. [5] quantified the overall amount of residual porosity of ceramics based on YAG using an optical microscope and found a relationship between the residual porosity and the scattering coefficient. In their study, they captured an image focused on each pore that they found in the analysed volume. This approach is effective in avoiding the "corpuscle problem" because the slicing process captures the pore at its maximum diameter and all the pores are recorded, not only the larger ones, thus simplifying the calculation. This method is particularly suitable for materials of high optical quality, in which the number of pores is low. However, it is important to note that the number of images quickly increases, making analysis slow.

In this study, we use a digital optical microscope for the rapid analysis of large sample volumes (~2.2×10^7 μm^3) and for a precise quantification of the number and size of pores, which are the main defects observed in YAG-based ceramics. The specific technique, called bulk defect analysis, is based on capturing focus-stacked images within the sample, which are then analysed using image analysis software. The depth of field of the optical microscope is much higher compared to CLSM and, above all, is higher than the distance between the stacked layers. This means that the image takes into account all the pores at their maximum diameter. Therefore, the "corpuscle problem" does not apply to bulk defect analysis in the same way as it does for Jin et al. [5].

For transparent materials, optical microscopy allows for the examination of not only the surface but also the entire volume of the sample. However, bulk defect analysis is not meant as a substitute for SEM, but as a complementary analysis. Bulk defect analysis may be particularly useful for the analysis of micrometric pores, which have greater scattering properties than nanopores (<150 nm) from the visible to NIR spectrum [12,24].

2. Experimental Methods

High-purity powders, including Y_2O_3 (≥99.99%, D_{50}: 0.22 μm; Nippon Yttrium Co., Ltd., Fukuoka, Japan), Al_2O_3 (≥99.99%, D_{50}: 0.12 μm; Baikowski, Poisy, France), and Yb_2O_3 (≥99.9%, D_{50}: 0.27 μm; Nippon Yttrium Co., Ltd., Fukuoka, Japan) were used. As sintering aids, TEOS (TetraEthyl OrthoSilicate, ≥99.999%, Sigma Aldrich, Burlington, MA, USA) and MgO (≥99.99%, D_{50}: 1.35 μm; Baikowski, Poisy, France) were incorporated. Defloc Z3 (Werner G. Smith Inc., Cleveland, OH, USA) was used as the dispersant at a concentration of 2 wt.% relative to the solids' weight. Samples with a composition of 10 at.% Yb:YAG ($Yb_{0.3}Y_{2.7}Al_5O_{12}$) were produced through reactive sintering in high vacuum furnace. The ceramic process is similar to the one reported in our previous work [25], with minor modifications, as reported in Table 1. The powders, TEOS, and the dispersant were dispersed in absolute ethanol with high purity (99.9%) alumina milling media in polyethylene bottles and mixed on a roller mill for 24 h at 80 rpm. The powder-to-ethanol ratio used was 1 to 2, while the powder-to-milling-media ratio was 1 to 2. Then, the milling media were separated from the suspension and the solvent removed via rotary evaporator or oven drying. The use of a rotary evaporator was expected to help avoid selective sedimentation of the powders since, during the desiccation, the suspension was kept in motion; after ethanol evaporation, the powders were placed in oven at 80 °C for 12 h. To remove the solvent via oven drying, the suspension is poured into a glass crystallization dish and placed into an oven at 80 °C for 24 h. The dried powders were sieved using a 134 μm (110 mesh) disposable plastic silkscreen. The samples were shaped into pellets with diameter of 15 mm via linear and isostatic pressing. For linear pressing, the powders were poured into a steel die and pressed at 80 MPa. The pressed pellets were cold-isostatic-pressed at 250 MPa. Pellets were calcinated in air at 800 °C for 1 h to remove humidity and residual organics. The calcinated pellets were sintered at 1750 °C for 16 h under high vacuum in a clean furnace with tungsten heating elements. In selected mixtures, an ultrasonication step was added between ball milling and desiccation using a 20 kHz ultrasonic probe (Sonics GEX500, Sonics & Materials, Inc., Newtown, CT, USA); the probe settings were amplitude of 35%, cycles of 10 s on and 2 s off, and total time of 20 min.

Table 1. List of samples and the relative differences in the preparation procedure.

Sample	TEOS	MgO	Dispersant	Ultrasonication	Desiccation
1	0.5 wt.%	-	no	no	rotary evaporator
2	0.5 wt.%	-	yes	no	rotary evaporator
3	0.5 wt.%	-	yes	yes	rotary evaporator
4	0.5 wt.%	-	yes	no	oven drying
5	0.5 wt.%	-	yes	yes	oven drying
6	0.5 wt.%	0.08 wt.%	yes	no	oven drying

The sintered samples underwent mirror polishing on both faces with diamond suspensions from 15 μm down to 0.25 μm using an automated polishing machine (Tegramin 25, Struers, Copenhagen, Denmark). Transmittance spectra were recorded with a UV-Vis spectrometer (Lambda 750, PerkinElmer Inc., Shelton, CT, USA) in the wavelength range from 200 to 1500 nm. The microstructure of the polished surfaces was analysed with a scanning electron microscope (ΣIGMA, Carl Zeiss microscopy GmbH, Oberkochen, Germany) equipped with Energy Dispersive X-ray analysis (EDX) probe INCA Energy X-Act (Oxford Instrument plc, Abingdon, UK). The X-ray diffractometry analysis (XRD) was performed with the D8 Advanced instrument (Bruker Co., Billerica, MA, USA).

In the bulk defect analysis, a digital optical microscope (RH-2000, Hirox Co., Tokyo, Japan) was employed. To conduct porosity analysis, an automated focus stacking function was used to acquire multiple images (slices) at different depths of focus and to merge them into a single image with only the in-focus pixels. The magnification used was 400×. The distance between each slice was 4 μm. Thanks to the high depth of field of the Hirox optical microscope (20 μm at 400×), all the pores in the focus-stacked image were captured at their maximum diameter. This enhanced the precision of the measurement and greatly simplified the calculations. The focus-stacked image is a flat image that represents a portion of the sample volume.

The acquired images were processed using the ImageJ (version 1.5.3) image analysis software. The software requires a binary (1 bit, black and white) image that distinguishes material from porosity. Image processing and analysis are shown in Figure 1. The analysed depth in the samples is 250 μm, and pictures are squares with a side length of approximately 300 μm. The total volume analysed with each picture is approximately 2.2×10^7 μm^3. Three pictures were captured in different areas of each sample. A spherical pore shape was considered.

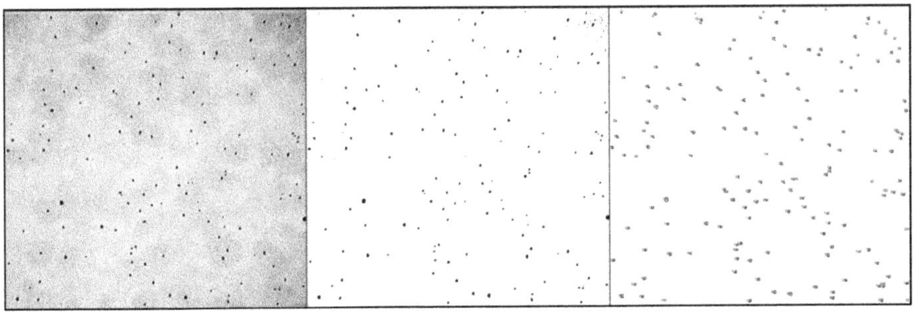

Figure 1. Focus-stacked image as captured (**left**), binary image (**centre**), and the ImageJ graphical output with the defects identified, highlighted, and numbered (**right**).

Using the obtained data, the average pore diameter, residual porosity by volume (rpV, the quantity generally described simply as porosity), and residual porosity by number (rpN, the number of pores per volume) were calculated using the following formulae:

$$rpV = \frac{\sum_1^n V_{pore}}{V_{analysed}} \times 10^6 \; [expressed\ in\ ppm] \tag{1}$$

$$rpN = \frac{total\ number\ of\ pores}{V_{analysed}} \times 10^6 \; [expressed\ in\ ppm] \tag{2}$$

3. Results and Discussion

Figure 2 displays photographs of the obtained samples; it can be observed that they all have a transparent appearance and are free from macro-defects.

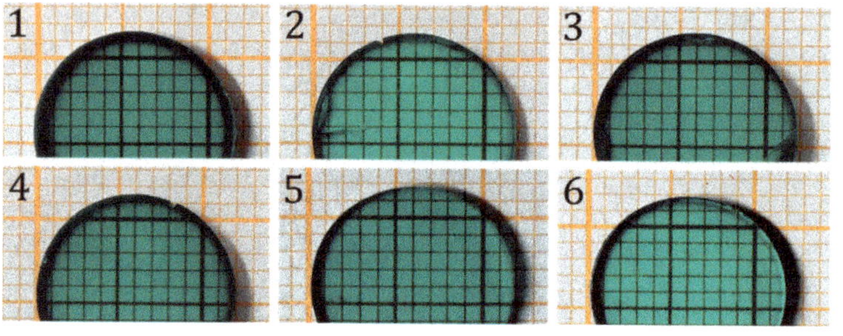

Figure 2. Photographs of analysed samples with their respective numbers as reported in Table 1; the green coloration is caused by the presence of Yb^{2+} ions.

The SEM analysis (Figures 3 and 4) revealed the absence of secondary phases. The identified defects comprised a few pores (highlighted by red circles in Figure 3), except for sample 3, where no defects were detected via SEM observation. However, the concentration of pores on the sample surface was too low to be accurately measured via SEM and to allow a comparison between samples. The composition and crystal structure of YAG were confirmed via EDX and XRD analysis (Figure 4).

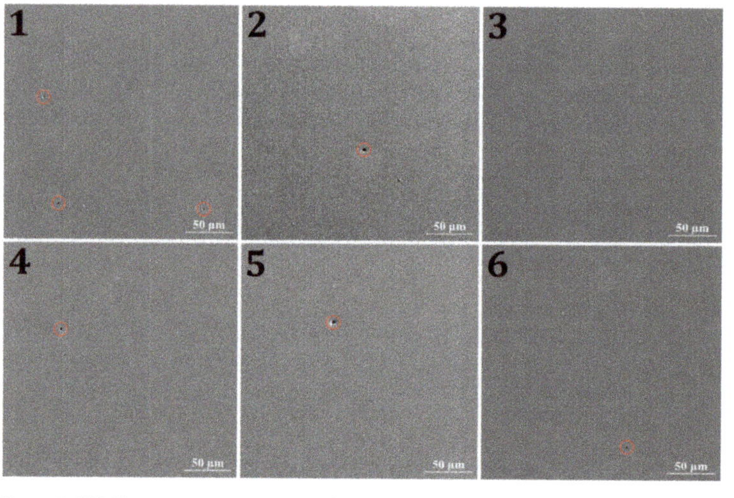

Figure 3. SEM images representative of the morphology of the polished surface of the samples with their respective numbers as reported in Table 1; red circles indicate pores.

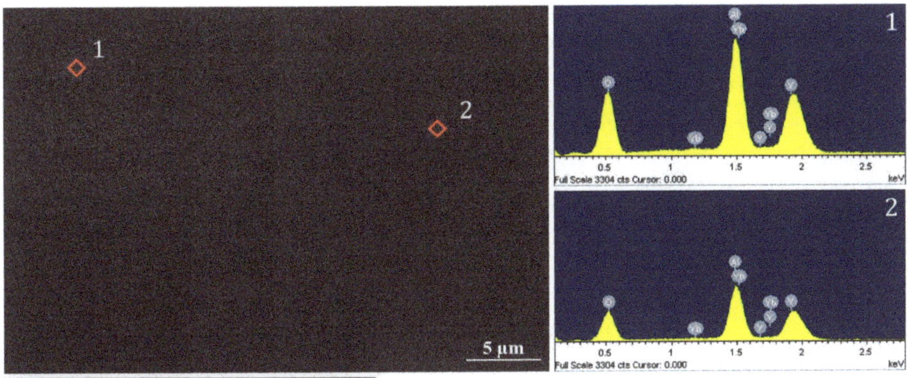

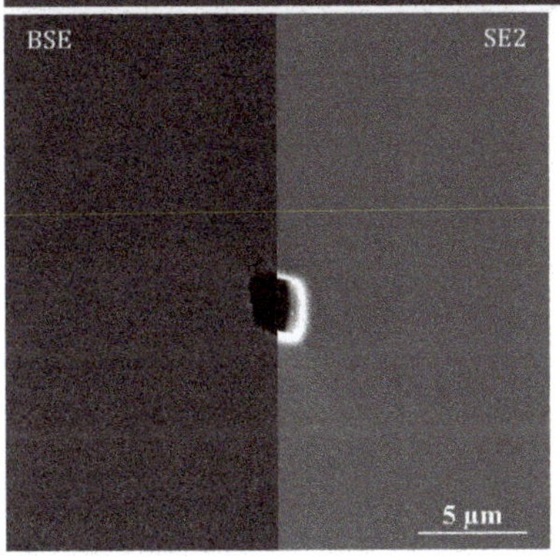

Element	Theorical [at. %]	1 [at. %]	2 [at. %]
O	60.00	63.18	62.58
Al	25.00	22.15	22.21
Y	13.50	12.99	13.61
Yb	1.50	1.67	1.60
Total	100.00	99.99	100.00

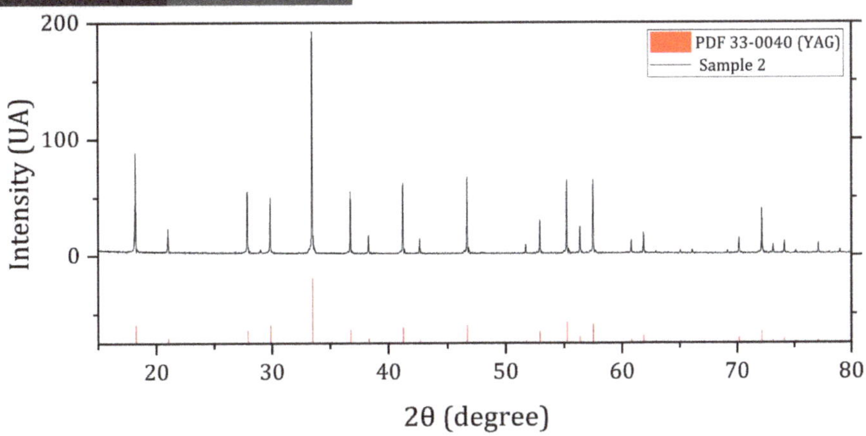

Figure 4. BSE image of a grain boundary and energy dispersive X-ray (EDX) analyses of two different grains in sample 1 ((**top**) and table); high-magnification SEM images of sample 1 near a pore with back-scattered electrons (BSE) and secondary electrons (SE2) detectors (**centre,left**); and XRD analysis of sample 2 (**bottom**).

Conversely, when we compared the transmittance spectra and the transmittance values at 1100 nm (Figure 5), differences emerged among the samples. Sample 6 exhibited the highest transmittance value, followed by samples 3 and 5. This suggested that despite the absence of visible pores on the surface of sample 3, there were pores within the volume. Sample 5 and 3 exhibited the same transmittance (79%, Table 2). On the other hand, sample

5 was considerably thicker than sample 3. As a consequence, sample 5 should have shown a lower overall porosity, i.e., lower linear losses, compared to sample 3.

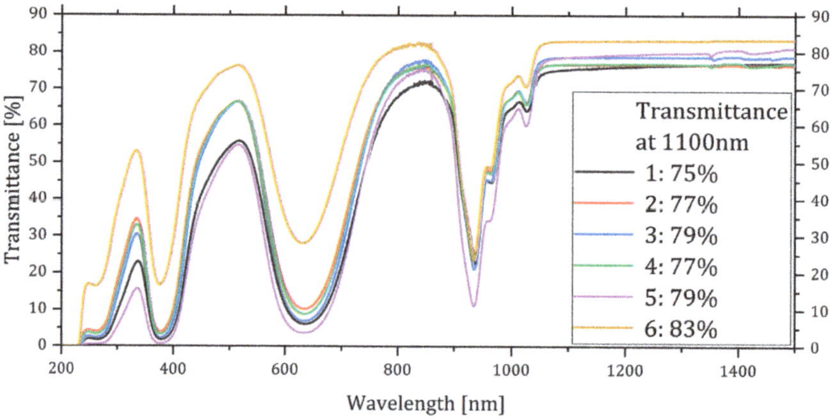

Figure 5. Transmittance spectra of the samples with their respective numbers as reported in Table 1 and value at 1100 nm.

Table 2. Results of the bulk defect analysis.

Sample	Thickness [mm]	Avg. Pore Size [μm]	*rpV* [ppm]	Transmittance @ 1100 nm
1	1.73	1.81 ± 1.09	66.7	75%
2	1.64	1.81 ± 1.21	25.4	77%
3	1.93	1.84 ± 1.09	27.4	79%
4	1.71	1.52 ± 1.29	24.0	77%
5	3.00	1.64 ± 1.13	21.0	79%
6	2.95	2.25 ± 1.69	13.6	83%

Further confirmation of the presence of defects within the volume was obtained by illuminating the samples from the edge, as shown in Figure 6. The key question was "how many defects are there?" The state-of-the-art techniques are not able to quantify the concentration of defects in the samples. Moreover, the size distribution of the defects cannot be derived from the transmittance curve, although the reverse is possible [12].

Figure 6. Sample 1 illuminated from the side; the presence of defects become evident.

On the contrary, the novel bulk defect analysis technique introduced here provided information on the size and concentration of defects, even when present in small amounts, as reported in Table 2 and Figure 7. This enabled a comprehensive understanding of the impact of various process parameters on the final samples' microstructure and properties. Since no secondary phases were detected via SEM, we assumed that all the defects observed with the optical microscope were pores. As shown in Table 2, the standard deviation of the pore size was quite high. On the other hand, the size of the pores varied considerably in each sample, as shown in Figure 7. Specifically, we found the following:

- samples 1 to 3, obtained via rotary evaporation, had a similar average pore size (around 1.8 μm);
- samples 4 and 5, obtained via oven drying, had a smaller pore size (around 1.6 μm) compared to 1, 2, and 3;
- sample 6, containing MgO, had the highest average pore size (2.25 μm).

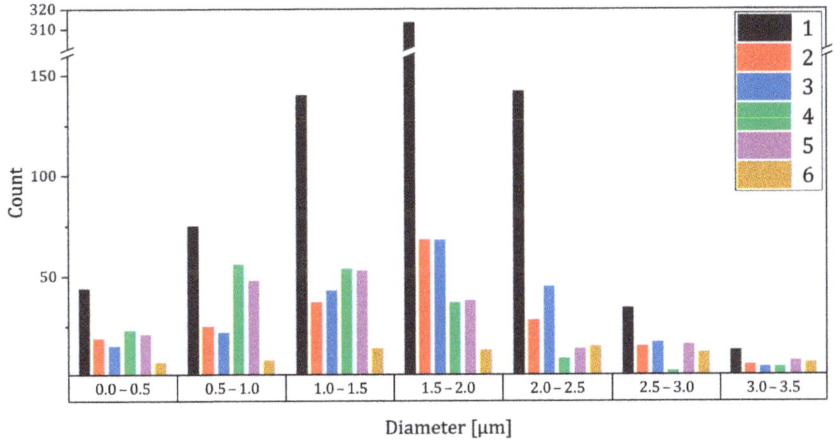

Figure 7. Pore size distribution histogram obtained via bulk defect analysis.

The variation in pore size distribution among the samples is also confirmed by the graph shown in Figure 6. Notably, the maximum number of pores for samples 1 to 3 is in the range from 1.5 to 2.0 μm, whereas for samples 4 and 5, it is in the range from 1.0 to 1.5 μm. The residual porosity by volume (*rpV*), i.e., the relative volume of the pores in the samples, shown in Table 2, decreases from sample 1 to 6, with the exception of sample 3. The highest *rpV* in sample 1 may be due to the absence of the dispersant in the mixing process. The dispersant helps to disperse the powders during milling and prevents reaggregation during drying. It also helps the particles to slide against each other during pressing, resulting in more homogenous powder packing. Samples 2 and 4 were prepared using a dispersant and differ in the drying step: rotary evaporation for sample 2 and oven drying for sample 4 (Table 2).

Samples 3 and 5 were also prepared using a dispersant and dried via rotary evaporation and oven drying, respectively, but the powder suspensions were ultrasonicated prior to drying. The ultrasonication probably increased the deagglomeration of the powder, as the transmittance was higher compared to samples 2 and 4. Ultrasonication did not seem to have a significant effect on the *rpV* or the average pore size, as shown in Table 2. Finally, sample 6 was prepared using a dispersing agent and a combination of TEOS and MgO as sintering aids and oven-dried. This sample had the highest transmittance of all the samples studied. On the other hand, it also had the highest average pore size combined with the lowest *rpV*.

For a more comprehensive comparison of samples with the highest transmittance (samples 3, 5, and 6), we also considered the residual porosity by number (*rpN*), i.e., the number of pores per unit volume, which has a direct impact on the scattering losses of transmittance.

As indicated in Table 3, the high transmittance of sample 6 is not only associated with a low *rpV*, but also a low *rpN* compared to samples 3 and 5. The very low pore concentration (*rpN*) is therefore the factor contributing to the higher transmittance, which conversely does not appear to be influenced by the pore size.

Table 3. Bulk defect analysis results for samples 3, 5, and 6 with the additional value of *rpN*.

Sample	Avg. Pore Size [µm]	*rpV* [ppm]	*rpN* [ppm]	Transmittance @ 1100 nm
3	1.84 ± 1.09	27.4	3.12	79%
5	1.64 ± 1.13	21.0	3.05	79%
6	2.25 ± 1.69	13.6	0.58	83%

4. Conclusions

In this study, bulk defect analysis, a technique based on digital optical microscopy, has been shown to be a method that is both fast and non-destructive, making it an effective way to quantify defects in the bulk of transparent ceramics, thanks to the fact that in the case of transparent materials, optical microscopy allows for the examination not only of the surface but also of the entire sample volume. Furthermore, this technique is capable of analysing larger volumes than other techniques.

In the research, the bulk defect analysis was used to provide detailed information on the average pore size and concentration of YAG-based ceramics doped with Yb. A volume of 2.2×10^7 µm^3 of each sample was analysed.

Upon visual inspection, it was observed that all the samples appeared transparent and free from macro-defects. However, differences between the samples were indicated by the transmittance spectra and the transmittance values at 1100 nm. The concentration of pores on the sample surface was too low to be accurately measured via SEM, making it difficult to compare between samples. bulk defect analysis revealed differences between the samples that were not visible via SEM. This provided a comprehensive understanding of how the process parameters affected the microstructure and properties of the final samples. In particular, it was observed that samples obtained from the rotary evaporator have a larger pore size compared to the ones obtained via oven drying. The combined use of MgO and TEOS resulted in a significant decrease in pore concentration in the material, even though an increase of pore size was observed. Additionally, the study found that there was a correlation between the increase in the transmittance value and the decrease in the *rpV* and *rpN* values.

Author Contributions: Conceptualization, F.P. and L.E.; methodology, F.P., J.H. and A.P.; investigation, F.P. and A.P.; resources, C.M.; writing—original draft preparation, F.P. and V.B.; writing—review and editing, V.B., J.H. and L.E.; supervision, J.H.; project administration, V.B.; funding acquisition, V.B. and L.E. All authors have read and agreed to the published version of the manuscript.

Funding: This research was funded by the Italian Ministry of Defence, grant number 8731, 4 December 2019 (CeMiLAP[2]).

Institutional Review Board Statement: Not applicable.

Informed Consent Statement: Not applicable.

Data Availability Statement: The raw data supporting the conclusions of this article will be made available by the authors on request.

Acknowledgments: The authors gratefully acknowledge support from the Italian Ministry of Defence u nder PNRM Contract No. 8731 of 4 December 2019 (CeMiLAP[2]). The authors gratefully acknowledge Werner G. Smith for the free supply of the dispersant Defloc Z3.

Conflicts of Interest: The authors declare no conflicts of interest.

References

1. Taira, T. Ceramic YAG lasers. *Comptes Rendus Phys.* **2007**, *8*, 138–152. [CrossRef]
2. Taira, T. RE^{3+}-Ion-Doped YAG Ceramic Lasers. *IEEE J. Sel. Top. Quantum Electron.* **2007**, *13*, 798–809. [CrossRef]
3. Ikesue, A.; Aung, Y.L.; Taira, T.; Kamimura, T.; Yoshida, K.; Messing, G.L. Progress in Ceramic LASERs. *Annu. Rev. Mater. Res.* **2006**, *36*, 397–429. [CrossRef]
4. Krell, A.; Klimke, J.; Hutzler, T. Transparent compact ceramics: Inherent physical issues. *Opt. Mater.* **2009**, *31*, 1144–1150. [CrossRef]
5. Jin, G.; Jiang, B.; Zeng, Y.; Li, J.; Liu, W.; Liu, J.; Pan, Y.; Sun, D. Study on the relation between optical scattering and porosity in transparent Nd/Yb:YAG ceramics. In *Pacific Rim Laser Damage 2011: Optical Materials for High Power Lasers*; Shao, J., Ed.; SPIE: Washington, DC, USA, 2011; p. 82061W.
6. Akinribide, O.J.; Mekgwe, G.N.; Akinwamide, S.O.; Gamaoun, F.; Abeykoon, C.; Johnson, O.T.; Olubambi, P.A. A review on optical properties and application of transparent ceramics. *J. Mater. Res. Technol.* **2022**, *21*, 712–738. [CrossRef]
7. Boulesteix, R.; Chevarin, C.; Belon, R.; Maître, A.; Cochain, L.; Sallé, C. Manufacturing of Large Size and Highly Transparent Nd:YAG Ceramics by Pressure Slip-Casting and Post-Sintering by HIP: An Experimental and Simulation Study. *Materials* **2020**, *13*, 2199. [CrossRef]
8. Hostaša, J.; Esposito, L.; Piancastelli, A. Influence of Yb and Si content on the sintering and phase changes of Yb:YAG laser ceramics. *J. Eur. Ceram. Soc.* **2012**, *32*, 2949–2956. [CrossRef]
9. Hostaša, J.; Picelli, F.; Hříbalová, S.; Nečina, V. Sintering aids, their role and behaviour in the production of transparent ceramics. *Open Ceram.* **2021**, *7*, 100137. [CrossRef]
10. Goldstein, A.; Krell, A. Transparent Ceramics at 50: Progress Made and Further Prospects. *J. Am. Ceram. Soc.* **2016**, *99*, 3173–3197. [CrossRef]
11. Xiao, Z.; Yu, S.; Li, Y.; Ruan, S.; Kong, L.B.; Huang, Q.; Huang, Z.; Zhou, K.; Su, H.; Yao, Z.; et al. Materials development and potential applications of transparent ceramics: A review. *Mater. Sci. Eng. R Rep.* **2020**, *139*, 100518. [CrossRef]
12. Hříbalová, S.; Pabst, W. Modeling light scattering by spherical pores for calculating the transmittance of transparent ceramics—All you need to know. *J. Eur. Ceram. Soc.* **2021**, *41*, 2169–2192. [CrossRef]
13. Esposito, L.; Piancastelli, A.; Costa, A.L.; Serantoni, M.; Toci, G.; Vannini, M. Experimental features affecting the transparency of YAG ceramics. *Opt. Mater.* **2011**, *33*, 713–721. [CrossRef]
14. Cavalli, E.; Esposito, L.; Hostaša, J.; Pedroni, M. Synthesis and optical spectroscopy of transparent YAG ceramics activated with Er^{3+}. *J. Eur. Ceram. Soc.* **2013**, *33*, 1425–1434. [CrossRef]
15. Lopez, A.R.; Baniel, P.; Gall-Borrut, P.; Granier, J.E. Scattering properties of ZrF4-based glasses prepared by the gas film levitation technique. In *Submolecular Glass Chemistry and Physics*; Bray, P., Kreidl, N.J., Eds.; SPIE: Washington, DC, USA, 1991; pp. 191–202.
16. Stuer, M.; Bowen, P.; Cantoni, M.; Pecharroman, C.; Zhao, Z. Nanopore Characterization and Optical Modeling of Transparent Polycrystalline Alumina. *Adv. Funct. Mater.* **2012**, *22*, 2303–2309. [CrossRef]
17. Boulesteix, R.; Maître, A.; Baumard, J.-F.; Rabinovitch, Y. Quantitative characterization of pores in transparent ceramics by coupling electron microscopy and confocal laser scanning microscopy. *Mater. Lett.* **2010**, *64*, 1854–1857. [CrossRef]
18. Kosyanov, D.Y.; Yavetskiy, R.P.; Parkhomenko, S.V.; Doroshenko, A.G.; Vorona, I.O.; Zavjalov, A.P.; Zakharenko, A.M.; Vornovskikh, A.A. A new method for calculating the residual porosity of transparent materials. *J. Alloys Compd.* **2019**, *781*, 892–897. [CrossRef]
19. Yavetskiy, R.P.; Doroshenko, A.G.; Parkhomenko, S.V.; Vorona, I.O.; Tolmachev, A.V.; Kosyanov, D.Y.; Vornovskikh, A.A.; Zakharenko, A.M.; Mayorov, V.Y.; Gheorghe, L.; et al. Microstructure evolution during reactive sintering of $Y_3Al_5O_{12}$:Nd^{3+} transparent ceramics: Influence of green body annealing. *J. Eur. Ceram. Soc.* **2019**, *39*, 3867–3875. [CrossRef]
20. Wicksell, S.D. The Corpuscle Problem. A Mathematical Study of a Biometric Problem. *Biometrika* **1925**, *17*, 84–99.
21. Saltykov, S.A. *Stereometric Metallography, (1958)*; Metallurgizdat: Moscow, Russia, 1979; p. 446.
22. Cruz Orive, L.M. Particle size-shape distributions: The general spheroid problem. *J. Microsc.* **1976**, *107*, 235–253. [CrossRef]
23. Pabst, W.; Uhlířová, T. A generalized class of transformation matrices for the reconstruction of sphere size distributions from section circle size distributions. *Ceram. Silik.* **2017**, *61*, 147–157. [CrossRef]
24. Zhang, W.; Lu, T.; Wei, N.; Wang, Y.; Ma, B.; Li, F.; Lu, Z.; Qi, J. Assessment of light scattering by pores in Nd:YAG transparent ceramics. *J. Alloys Compd.* **2012**, *520*, 36–41. [CrossRef]
25. Picelli, F.; Biasini, V.; Hostaša, J.; Piancastelli, A.; Esposito, L. A useful approach to understand the origin of defects in transparent YAG ceramics. *MRS Commun.* **2022**, *12*, 807–812. [CrossRef]

ceramics

MDPI

Article

Optical and Spectroscopic Properties of Ho:Lu_2O_3 Transparent Ceramics Elaborated by Spark Plasma Sintering

Lucas Viers [1], Simon Guené-Girard [2], Gilles Dalla-Barba [3], Véronique Jubéra [2], Éric Cormier [3], Rémy Boulesteix [1,*] and Alexandre Maître [1]

1 Faculty of Science and Technology, University of Limoges, CNRS, IRCER, UMR 7315, F-87068 Limoges, France
2 Bordeaux INP, University of Bordeaux, CNRS, ICMCB, UMR 5026, F-33600 Pessac, France
3 Institut d'Optique Graduate School, University of Bordeaux, CNRS, LP2N, UMR 5298, F-33400 Talence, France
* Correspondence: remy.boulesteix@unilim.fr

Abstract: In this work, transparent ceramics were manufactured from nanopowders synthesized by aqueous coprecipitation followed by Spark Plasma Sintering (SPS) to ensure rapid and full densification. The photoluminescence of Ho:Lu_2O_3 transparent ceramics was studied in the Visible and IR domains as a function of Ho^{3+} dopant level from 0.5 at.% to 10 at.%. A cross-relaxation mechanism was identified and favors the 2 μm emission. All of the obtained results indicate that the optical properties are very similar between $Lu_{2-x}Ho_xO_3$ transparent ceramics and single crystals. Thus, the SPS technique appears to be a very promising method to manufacture such ceramics, which could be used as amplifier media for high-energy solid-state lasers.

Keywords: Lu_2O_3; holmium; transparent ceramics; photoluminescence; lifetime; SPS

1. Introduction

In recent years, transparent ceramics have appeared as a very interesting new class of material for high-power solid-state lasers, leading to the development of multi-kilowatt laser systems [1–4]. Among the variety of the targeted laser characteristics, lasers emitting at 2 μm have been mainly studied for their great potential in the medical field because of the absorption bands of water around this wavelength [5–8]. Compared to other 2 μm emitting ions, Ho^{3+} has the advantage of presenting the largest emission cross-section [9]. Due to their high mechanical resistance, good chemical stability and high thermal conductivity, rare-earth sesquioxides (Sc_2O_3, Y_2O_3 and Lu_2O_3) and garnet structures like YAG (Yttrium Aluminum Garnet) are well adapted and the most used host matrix for such ions [10–13]. Among these materials, Lu_2O_3 is of primary interest because it maintains a high thermal conductivity close to 12 $W{\cdot}m^{-1}{\cdot}K^{-1}$ even for high concentrations of dopant [14].

A high concentration of active ion has great potential for use in high-power laser applications, especially in sesquioxide ceramic matrices where the dopant is inserted in a solid solution [15,16]. Indeed, it can be supposed that a higher doping rate could induce a higher luminescence intensity [17,18]. However, a high dopant volume density can also generate undesired energy transfers, called non-radiative transitions, such as cross-relaxation (CR) or excited-state absorption (ESA), causing quenching of the photoluminescence signal. For the Ho^{3+} ion, the most common transfers are CRs between two ions, called the donor and acceptor, which are performed by up-conversion or down-conversion depending on the initial state of the acceptor and the final state of the donor [19].

The main results concerning the previous developments of sesquioxide ceramics for laser applications were recently summarized in the review of Liu et al. [20]. According to their conclusions, Lu_2O_3 sesquioxide ceramics seem to be the most promising host materials for high-power laser operation. However, they highlight the long road ahead to obtain transparent sesquioxide ceramics of large size and high optical quality. Especially, they pointed out the crucial role played by the synthesis step of nanopowders, as well as the production of powder compacts with ideal microstructures. According to the works cited in the literature, the elaboration of transparent ceramics requires control of each process step, especially that of sintering. In fact, full densification and total elimination of porosity is mandatory to ensure high optical properties and high laser performances. Pressureless sintering under vacuum, pressure-assisted densification or a combination of these techniques is an efficient way to remove all pores, and non-conventional sintering techniques like Hot Pressing (HP), Hot Isostatic Pressing (HIP) or Spark Plasma Sintering (SPS also called Field Assisted Sintering Technique—FAST) have been developed and extensively studied for a wide variety of ceramics. Some recent studies were lead, for example, on Y_2O_3 by HP + HIP [21], Yb:Lu_2O_3 and Tm:Lu_2O_3 by vacuum sintering + HIP [22,23] or Er:$(Sc_xY_{1-x})_2O_3$ by vacuum sintering [24]. Among other techniques, Spark Plasma Sintering relies on the densification of a powder inside an electrically conductive matrix, heated by the Joule effect thanks to the application of a pulsed current with a high intensity, and submitted to a uniaxial pressure generally in the range of 50–150 MPa. SPS is increasingly used for fast and efficient densification of numerous ceramics, including nitrides, carbides and oxides [25]. However, the fast shrinkage and physical phenomena involved during sintering lead to a lack of knowledge on sintering mechanisms, more precisely on the densification step and the subsequently formed microstructures. Moreover, carbon contamination from the common use of a graphite matrix is a main issue often encountered for the sintering of oxides and could strongly affect their optical and spectroscopic properties [26]. Concerning the elaboration of transparent ceramics, few studies have compared the microstructure optical properties' relationships to other non-conventional sintering techniques or have highlighted the specificities of SPS on this family of materials. As an example, some studies were interested in Nd:Lu_2O_3 [27,28] or Yb:Sc_2O_3 [29] ceramics elaborated by either SPS or HIP techniques. Their main conclusions highlighted that such ceramics in the rare-earth oxide family have similar optical properties to single crystals. In the case of the Ho^{3+} ion, no study has been reported on the implication of the fabrication process parameters on the luminescence properties of as-obtained materials, more especially on the CR mechanisms.

In this work, transparent ceramics of Lu_2O_3 with various Ho^{3+} doping rates were manufactured by a process combining nanopowder synthesis and direct densification by Spark Plasma Sintering. Chemical synthesis by the co-precipitation route has the advantage of producing well-crystallized Ho-doped Lu_2O_3 nanopowders with nanometric grain size and great chemical homogeneity [30]. The microstructure (i.e., grain size), luminescence properties and optical transparency of as-obtained transparent ceramics were studied as a function of the holmium doping content. The results were compared to ones reported in the literature for similar single crystals or transparent ceramics manufactured by HIP in order to highlight the specificities induced by the SPS process.

2. Materials and Methods

2.1. Synthesis of Ho:Lu_2O_3 Nanopowders

Ho-doped Lu_2O_3 nanopowders were synthesized by inverse co-precipitation. Rare-earth nitrates were used as precursor and ammonium bicarbonate as precipitant. Commercial oxides of Lu_2O_3 and Ho_2O_3 (AUER REMY, Lehvoss group, Hamburg, Germany) were first dissolved in hot nitric acid (HNO_3, 65 vol.%, Thermo Fisher Scientific, Waltham, MA, USA). The nitrate solution was poured drop by drop at a rate of 3 mL·min^{-1} into the solution of ammonium bicarbonate (NH_4HCO_3, 99wt.%, Acros Organics, Thermo Fisher Scientific, Waltham, MA, USA). The co-precipitation was operated at room temperature with vortex stirring according to the set-up described in Figure 1.

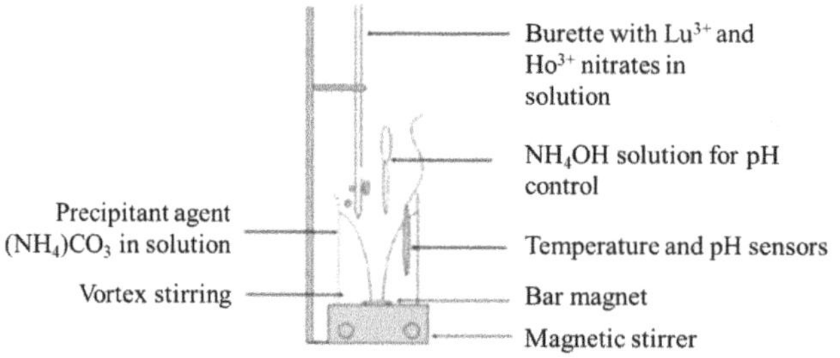

Figure 1. Set-up used for $Ho:Lu_2O_3$ precipitation.

Ammonia solution (NH_4OH, 35 vol.%, Thermo Fisher Scientific, Waltham, MA, USA) was added to maintain the pH at 9. Then, the solution was aged for 24 h under stirring at room temperature. The obtained precipitate was washed three times with distilled water by centrifugation and three times with ethanol and was dried at 70 °C. The dried precipitate was milled in an agate mortar and calcined at 1000 °C for 2 h under air to form the final crystallized $Ho:Lu_2O_3$ nanopowder. Powder morphology, structure and chemical homogeneity were obtained by TEM (Transmission Electronic Microscopy) in classical or scanning (STEM) modes and high-resolution TEM coupled with SAED (Selected Area Electron Diffraction) (2100 F, JEOL, Tokyo, Japan). Crystalline phases were identified by X-ray diffraction analyses (D8, Bruker, Karmsruhe, Germany) using CuKα radiation. CuKα2 radiation contribution was removed and obtained diagrams were indexed with DIFFRACplus EVA™ software V5 and the PDFmain™ database.

The TEM observations reported in Figure 2a show a slightly edged shape and narrow size distribution of particles with a mean diameter of about 63 nm. No holmium segregation in the form of isolated Ho_2O_3 particles was detected, as illustrated by the STEM micrograph in Figure 2b. That means there is no differential precipitation of Ho and Lu cations during the synthesis process. The particles also appeared well crystallized in the bixbyite structure (Ia-3, cubic C-type), as revealed by diffraction pattern (Figure 2c). As a result, the $Ho:Lu_2O_3$ nanopowder can be considered as a $Lu_{(2-x)}Ho_xO_3$ homogeneous solid-solution ($x = 0.02$ for 1 at.% holmium doping content) with a cubic crystalline structure. The diffraction patterns registered as a function of the calcination temperature in Figure 2d also prove that the powder is single phased in the bixbyite structure and well crystallized for temperatures higher than 600 °C. A calcination temperature of 1000 °C was chosen regarding the higher crystallinity of the powder particles at this temperature.

2.2. *Densification by Spark Plasma Sintering*

A commercial device (825 series 8000 A, FUJI-SPS, Saitama, Japan) was used for SPS experiments. The powder samples were put in a 13 mm diameter graphite die with Papyex® foil (MERSEN, La Défense, France). Sintering experiments were led under primary vacuum ($P < 20$ Pa) with a pulse sequence of 12:2. A digital IR pyrometer was used for thermal regulation on the surface of the graphite die, which started detecting at a minimum temperature of 573 °C. For all treatments used to elaborate fully dense ceramics, the heating rate was fixed to 100 $°C{\cdot}min^{-1}$ until 1200 °C. Then, the heating rate was 10 $°C{\cdot}min^{-1}$ until 1400 °C, where the temperature was maintained for 15 min under uniaxial pressure of 130 MPa. The cooling rate was 100 $°C{\cdot}min^{-1}$. Then, ceramics were annealed under air at 985 °C for at least 50 h to re-oxidize them and eliminate the carbon contamination caused by the contact with graphite.

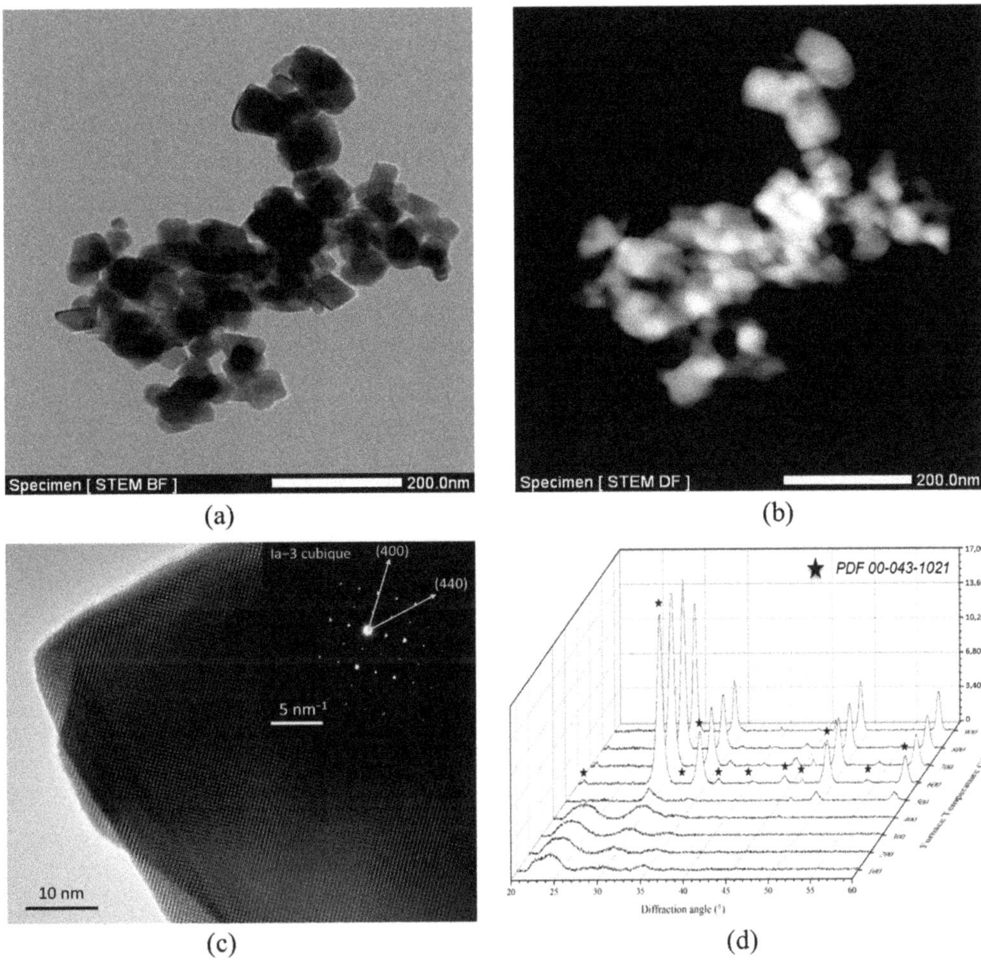

Figure 2. STEM micrographs of the coprecipitated 1 at.%Ho:Lu_2O_3 powder calcined at 1000 °C in bright field (**a**) or in dark field (**b**), and high-resolution TEM micrograph with corresponding diffraction pattern (**c**). XRD patterns registered as a function of the calcination temperature of 1 at.%Ho:Lu_2O_3 powder (**d**).

2.3. Characterization of Sintered Samples

Prior to optical characterizations, the samples were ground to obtain flat and parallel surfaces and then mirror polished with 30 nm colloidal silica after pre-polishing with SiC papers and diamond pastes. A Cary 5000 spectrophotometer was used to record the transmission curves of transparent samples (Agilent, Santa Clara, CA, USA). The spectra resolution was 1 nm with an integration time equal to 0.1 s.

The photoluminescence spectra of various doped ceramics were obtained from 450 nm to 2200 nm into 3 sub-intervals: Visible [450–800 nm], Near-IR [1000–1300 nm] and Mid-IR [1800–2200 nm]. For all ranges, a photoluminescence spectrometer was used (Fluorolog3, Horiba, Kyoto, Japan) using a 450 W Xe lamp for excitation in Visible and 450 nm laser diode (50 mW) for Near-IR and Mid-IR. For detection, a photo multiplicator detector (PMT-Hamamatsu R298, Shizuoka, Japan) was used in Visible, an InGaAs detector for Near-IR and a PbS photoconductor for Mid-IR. For Visible acquisitions, the spectral resolution was 2 nm/mm of opening slits (for both entry and exit ones). Some corrections were applied to the raw spectra to account for variations in the intensity of the excitation source, set-up geometry and detector sensitivity. For measurements in IR range, the spectral resolution was 12 nm/mm of opening slits and the integration time was 0.5 s.

Lifetime measurements were performed with a 10 nm step-by for each wavelength using an Optical Parametric Oscillator (OPO) for source pumping. Infrared lifetime measurements were performed with an Edinburgh FLS1000 spectrometer coupled with a digital Tektronix oscilloscope to record the time-dependent decay. All the measurements were taken at room temperature.

3. Results

3.1. Sample Microstructure and Optical Transmittance

After sintering at 1400 °C for 15 min, annealing and polishing, Lu_2O_3 ceramics with doping rates from 0.5 at.% to 10 at.% Ho were obtained (Figure 3). All specimens were 2 mm thick and presented good transparency.

Figure 3. Transparent Ho:Lu_2O_3 ceramics with various Ho^{3+} doping rates (from the left to the right: 0.5 at.%, 1 at.%, 5 at.% and 10 at.%) sintered by SPS at 1400 °C for 15 min under 130 MPa of uniaxial pressure.

The SEM micrographs taken for the 1 at.%Ho:Lu_2O_3 sample (Figure 4a) showed a dense microstructure with a small and homogeneous grain size in the order of 200–500 nm. Nevertheless, few intergranular and/or intragranular residual pores were also found, especially by the TEM observations (Figure 4b). Their diameter did not exceed 50 nm and their concentration was estimated to be 90 pores·μm^{-3}. These defects could be related to the presence of microstructural inhomogeneities in powder compacts before sintering and/or rapid grain growth [31]. In fact, it has been highlighted in a previous paper that grain growth starts to be significant for temperatures higher than 1300 °C in such ceramics during sintering by SPS, independently of the applied pressure [32].

The transmission curves of the samples are presented on Figure 5. Data have been obtained for wavelengths from 400 nm to 2500 nm. All of the ceramics present the same optical absorption peaks corresponding to 4f electronic transitions of Ho^{3+} ion with corresponding intensity generally increasing with the doping rate. The most intense absorption peaks between 1800 nm and 2200 nm correspond to the $^5I_8 \rightarrow ^5I_7$ transition level. The optical transmittance baseline of all samples presents a continuous decrease from wavelengths of 1200 nm to 300 nm. The obtained curves were compared to the maximum theoretical transmittance of the Lu_2O_3 samples. To determine these values, one has to consider the general law giving the transmittance of a transparent material (Equation (1)):

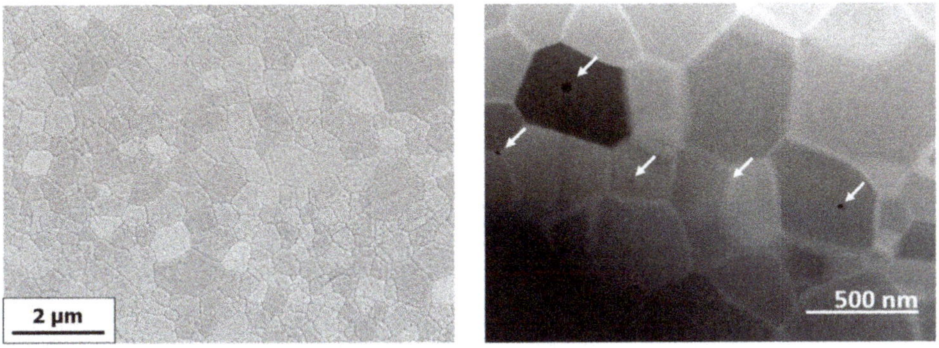

(a) (b)

Figure 4. SEM (**a**) and TEM (**b**) micrographs of transparent 1 at.%Ho:Lu_2O_3 ceramics sintered by SPS at 1400 °C for 15 min under 130 MPa of uniaxial pressure. Residual porosity was identified by TEM (white arrows).

$$T(\lambda) = T_{max}\exp(-\alpha x) \tag{1}$$

where T_{max} is the maximal transmittance that could be obtained for a perfectly polished sample with planar and parallel surfaces, α is the attenuation coefficient and x is the sample thickness. By considering multiple reflections at the sample's surface (no antireflective coatings were applied in our case), the maximal transmittance at normal incidence is given by Equation (2):

$$T_{max}(\lambda) = \frac{2n(\lambda)}{\left(1 + n(\lambda)^2\right)} \tag{2}$$

where $n(\lambda)$ is the refractive index of cubic Lu_2O_3 given by Equation (3) [33]:

$$n(\lambda)^2 = 3.6196 + \frac{0.04131}{\lambda^2 - 0.0238} - 0.00856 \cdot \lambda^2 \tag{3}$$

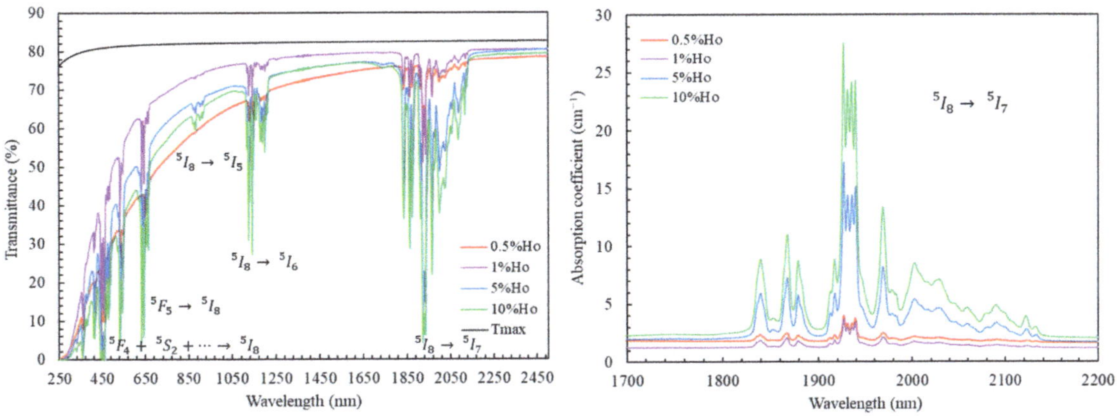

(a) (b)

Figure 5. Optical transmission (**a**) and absorption (**b**) as a function of wavelength of Ho:Lu_2O_3 transparent ceramics with various Ho^{3+} doping rates.

As a result, one is able to calculate the maximum transmittance value for a perfectly transparent Lu_2O_3 sample. The values are reported in Figure 5.

According to Figure 5a, a strong difference appears between the maximum transmittance and the transmittance baseline of Lu_2O_3 samples. This phenomenon is well-known for transparent ceramics and is generally induced by light scattering coming from residual porosity [34]. However, despite the presence of such optical losses, all Ho-doped Lu_2O_3 ceramics have a transmittance higher than 75%, i.e., more than 90% of the theoretical limit, for wavelengths > 2 µm. On the absorption curves (Figure 5b), the effect of the doping rate is clearly highlighted. Indeed, the intensity of absorption transition $^5I_8 \rightarrow {}^5I_7$ increases gradually with the Ho^{3+} concentration.

The absorption cross-sections of the Ho:Lu_2O_3 ceramics σ_{abs} were then calculated for all dopant contents from linear absorption coefficient α_λ and by using Equations (4) and (5):

$$\alpha_\lambda = (A_\lambda / x) \cdot \ln(10) \quad (4)$$

$$\sigma_{abs} = \alpha_\lambda / C_{Ho} \quad (5)$$

where A_λ is the absorbance for a specific wavelength λ, x is the thickness of sample and C_{Ho} is the volume density of Ho^{3+}. The values of volumetric density of Ho^{3+} (Table 1) were determined thanks to Equation (3) below:

$$C_{Ho} = (\rho \cdot \%at_{Ho} \cdot N_A) / (M_{Lu2O3:Ho}) \quad (6)$$

where ρ is the theoretical volumetric mass of Lu_2O_3 with Ia-3 cubic crystalline structure (obtained by XRD measurements), $\%at_{Ho}$ is the doping rate, N_A is the Avogadro constant and $M_{Lu2O3:Ho}$ is the molar mass of $Lu_{2-x}Ho_xO_3$ compound at a specific doping rate (0.01 < x < 0.2).

Table 1. Spectroscopic characteristics of $Lu_{2\text{-}x}Ho_xO_3$ samples as a function of Ho^{3+} concentration.

%at. Ho	0.5%	1%	5%	10%
C_{Ho} (cm^{-3})	1.43×10^{20}	2.85×10^{20}	1.42×10^{21}	2.84×10^{21}
Absorption cross section at 1927 nm (cm^2)	1.26×10^{-20}	1.03×10^{-20}	1.26×10^{-20}	1.03×10^{-20}
Fluorescence lifetime at 2030 nm (ms)	-	11.5	-	4.92

The data reported in Table 1 were used to plot the evolution of the absorption cross-section in the wavelength range 1900–1960 nm for all samples (Figure 6). It was shown that the curves seemed to follow the same evolution of σ_{abs} whatever the doping rate. Moreover, they overlapped despite a slight offset for 0.5% and 5%, which was obviously caused by a lower transmittance baseline (i.e., higher scattering by residual porosity) obtained for both samples. However, for the samples with low residual scattering, the calculated value of σ_{abs} was approximately 1×10^{-20} cm^2 at 1927 nm, which corresponded to the measured data for some Ho-doped systems, for example Ho:YAG [35].

3.2. *Photoluminescence and Fluorescence Decay Time*

To highlight the different energy transfers and associated light emissions and to attempt to locate some specificities of the elaboration process by SPS, photoluminescence spectroscopy was carried out on all samples in three interesting wavelength domains to highlight potential CR transitions: Visible (from 450 to 800 nm), Near-IR (from 1000 to 1300 nm) and Mid-IR (from 1800 to 2200 nm). The as-obtained results are summarized in Figure 7.

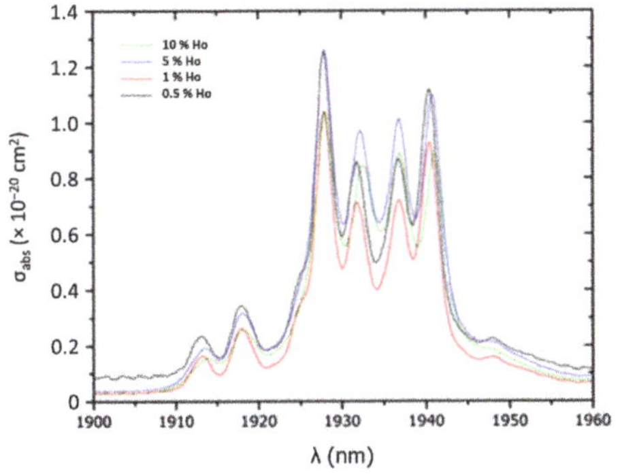

Figure 6. Evolution of absorption cross-section of Ho:Lu_2O_3 transparent ceramics with various Ho^{3+} doping rates.

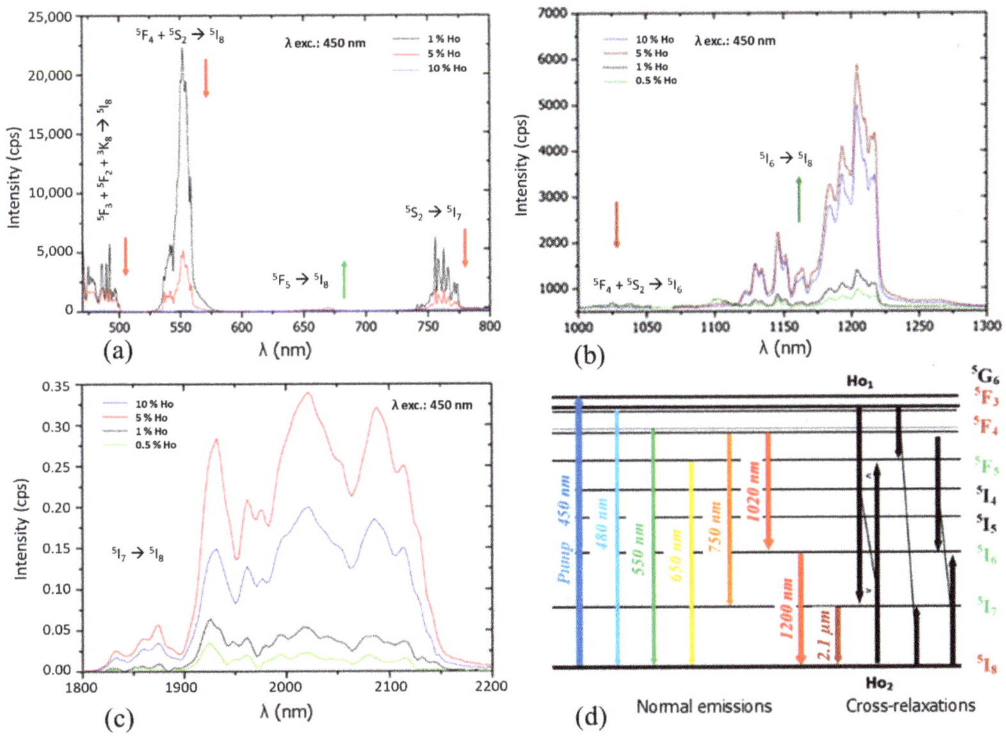

Figure 7. Emission spectra (excitation at 450 nm) of Ho:Lu_2O_3 transparent ceramics in the Visible range (**a**), Near-IR range (**b**) and Mid-IR (**c**) and the associated energy levels diagram built from experimentation (**d**).

In the Visible range, the emission bands symbolized by red arrows presented a strong decrease when the doping rate increased, with an extinction at 10 at.%Ho (Figure 7a). The identified transitions were $^5F_3 + {^5F_2} + {^3K_8} \rightarrow {^5I_8}$ at 480 nm, $^5F_4 + {^5S_2} \rightarrow {^5I_8}$ at 550 nm and $^5S_2 \rightarrow {^5I_7}$ at 750 nm. However, the transition at 650 nm corresponding to $^5F_5 \rightarrow {^5I_7}$ was the only one to increase when the Ho^{3+} content increased. This change in the relative lines intensity was also observed in the Near-IR range, where the transition $^5F_4 + {^5S_2} \rightarrow {^5I_6}$ at 1020 nm decreased, whereas the one at 1200 nm, associated with $^5I_6 \rightarrow {^5I_8}$, increased until Holmium doping reached 5%. A maximum of the $^5I_7 \rightarrow ^5I_8$ transition in Mid-IR at 1.8–2.1 μm was observed for the same Ho^{3+} content (Figure 7b,c).

These variations in the intensities of the emission bands are generally representative of non-radiative transitions occurring in the system when the doping rate increases [19,36]. Indeed, when the dopant volume density increases, the average distance between the two Ho^{3+} ions is smaller, which facilitates undesired interactions and the generation of a concentration quenching phenomenon. In addition to this loss of emission intensity due to energy transfer on defects, it is also possible to observe some possible mechanisms corresponding to Cross-Relaxation (CR) energy transfers described in Figure 7d [37], depending on the Ho^{3+} content:

(1) At a low doping rate (<1%), and after excitation according to the 450 nm transition $^5I_8 \rightarrow ^5G_6$, a cascade of transitions occurs. First, $^5F_3 + ^5F_2 + ^3K_8 \rightarrow ^5I_8$ radiative transition at 480 nm happens, as well as a non-radiative $^5G_6 \rightarrow ^5F_4 + ^5S_2$ transition. From this electronic level, two optical emissions occur at 550 nm and 750 nm corresponding, respectively, to the $^5F_4 + ^5S_2 \rightarrow ^5I_8$ and $^5F_4 + ^5S_2 \rightarrow ^5I_7$ transitions. In a second step, the transition $^5I_7 \rightarrow ^5I_8$ at 2100 nm occurs. From level $^5F_4$, $^5S_2$, a second transition $^5F_4 + ^5S_2 \rightarrow ^5I_6$ at 1020 nm, as well as a non-radiative transition $^5F_4 + ^5S_2 \rightarrow ^5F_5$, is observed. Then, from the $^5F_5$ electronic level, an emission is present at 650 nm, linked to the $^5F_5 \rightarrow ^5I_8$ transition.

(2) When the doping increases, due to the increase in the volume concentration of Ho^{3+} ions in the matrix, Ho_1-Ho_2 non-radiative Cross-Relaxation interactions are increasingly favored. These interactions, which induce modification of electronic level populations, lead to the overcrowding of the $^5I_7$ and $^5F_5$ levels according to the $^5I_8 \rightarrow ^5I_7$ or $^5F_3 + ^5F_2 \rightarrow ^5I_7$ transitions of Ho_1 and $^5F_3 + ^5F_2 \rightarrow ^5F_5$ or $^5I_8 \rightarrow ^5F_5$ of Ho_2. Thus, the intensity of the $^5I_7 \rightarrow ^5I_8$ transition in the Mid-IR at 2100 nm increases at the expense of Visible transitions up to 5% in Ho^{3+} ions (Figure 7a,c). Finally, a last relaxation takes place and induces the overcrowding of level $^5I_6$ according to the transitions $^5F_4 + ^5S_2 \rightarrow ^5I_6$ and $^5I_8 \rightarrow ^5I_6$. Hence, the transition in the Near-IR $^5I_6 \rightarrow ^5I_8$ at 1200 μm is favored (Figure 7b).

(3) Above 5%, the phenomenon of fluorescence quenching by concentration is observed for all transitions [38]. For the highest concentration tested, namely 10% (2.84×10^{21} ions·cm^{-3}), the fluorescence intensity in the Near-IR and Mid-IR is reduced by 40% compared to 5% (1.42×10^{21} ions·cm^{-3}) (Figure 8b,c).

Figure 9 shows the fluorescence lifetime measured on the two extreme doping rates (i.e., 1% and 10%) in order to see the influence of Ho^{3+} concentration on the fluorescence decay of $^5I_7 \rightarrow ^5I_8$ laser transition at 2100 nm. Data were obtained for an excitation at 1850 nm and an emission at 2030 nm, corresponding to the maximum intensity observed in Figure 7c. The fluorescence lifetime was obtained by curve fitting with an exponential function. For 1%, a lifetime of 11.5 ms was obtained for $^5I_7$ level state. Then, this lifetime decreased from 11.5 ms to 4.92 ms when the doping rate increased from 1% to 10%, which is in accordance with the literature on Ho-doped sesquioxides [39].

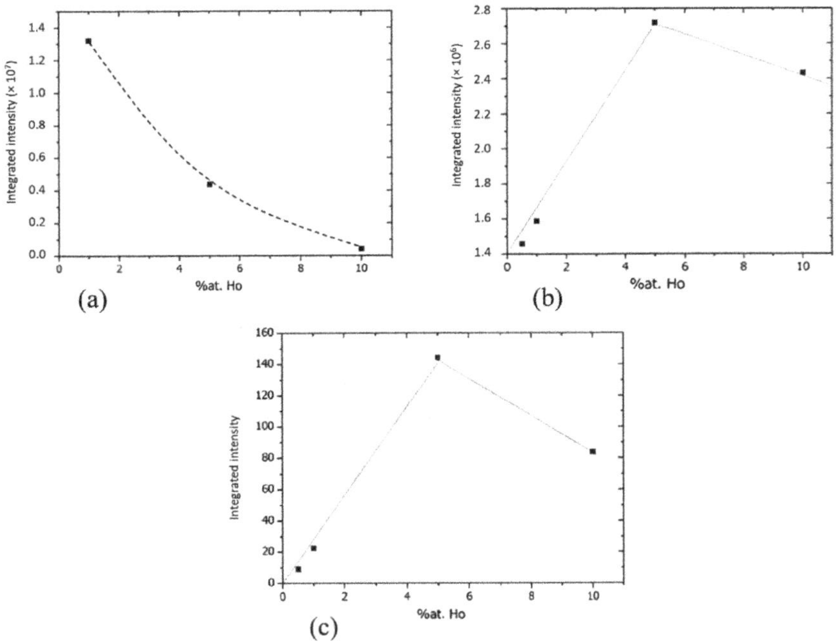

Figure 8. Evolution of integrated emission intensities of Ho:Lu_2O_3 transparent ceramics in the Visible (400–800 nm) range, (**a**) Near-IR range (1000–1300 nm) (**b**) and Mid-IR range (1800–2200 nm) (c).

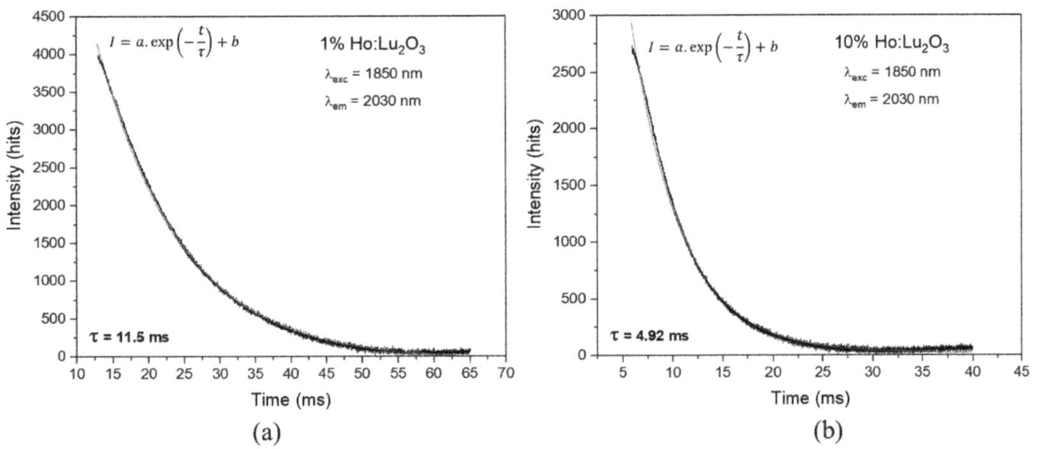

Figure 9. Fluorescence lifetime of 1% (**a**) and 10% (**b**) transparent ceramics. Fitted curves are represented in red lines with associated exponential decay function.

This phenomenon confirms the concentration quenching of the 2.1 µm emission caused by the high active ion density inside material, as mentioned in Section 3.1.

4. Discussion

In this work, transparent $Lu_{2-x}Ho_xO_3$ ceramics with fine microstructures (G < 1 µm) were obtained thanks to the SPS sintering technique. It increases the list of Lu_2O_3-based transparent ceramics obtained by SPS, previously doped with Nd [27,28], Tm [13] or Yb [40], proving the capability and versatility of this technique. Some of them have proven sufficient optical quality to observe a laser oscillation. One can cite also the works of Toci et al., who first demonstrated a laser effect with a 1 at.% $Nd:Lu_2O_3$ transparent ceramic obtained by SPS [41]. A few years later, Xu et al. achieved a laser slope efficiency of 38% with a maximum output of 1.25 W [42]. Many other available sintering techniques have been reported in the literature, such as air, vacuum or hot pressing (HP) pre-sintering followed by high-temperature post-HIP for a long time. The main drawback in this case relies on the large-sized grains (G > 10 μm), which are detrimental to the thermal and mechanical properties of ceramics. As an example, Kim et al. obtained HP + HIP Ho^{3+}-doped Lu_2O_3 ceramics with an average grain size around 40–50 μm [39]. This difference is a consequence of the thermomechanical cycle used during the SPS treatment, performed at lower maximum temperatures and faster heating rates, which drastically limit grain growth during sintering [32]. As a result, compared to HIP, some differences in the optical properties, originating from the difference in the microstructure or carbon contamination of the samples after SPS treatment, could be expected.

First, concerning the effect of grain size, it seems to have no significant effect for Ho-doped Lu_2O_3. Indeed, by comparing data to 2% Ho-doped Lu_2O_3 ceramic obtained by HIP [39], a similar fluorescence lifetime in the order of 10 ms is obtained at a wavelength of 2140 nm. This value is also of the same order of magnitude as that obtained for single crystals of similar composition. The mean grain diameter obtained in this work is around 50 times lower than that obtained by HIP. That means that the volume of grains is about $(G/G')^3$ ~10^5 lower in our samples and the grain boundaries volume fraction is about (G′/G)~50 times higher. Therefore, the environment of Ho at grain boundaries should be similar than that observed in the bulk, and very limited Ho segregation is expected to occur in such material. This shows an advantage of this system for laser applications compared to other ones where dopant segregation at grain boundaries is observed, like in rare-earth-doped fluoride ceramics [43].

Second, concerning the use of SPS treatment, it seems to have no observable effect on the photoluminescence properties or lifetime decay according to results reported in Table 2. For example, on a 1%$Ho:Lu_2O_3$ single crystal, Dong et al. obtained a similar value of 9.8 ms for the same optical transition at a wavelength of 2030 nm [44]. A similar trend was observed for a lower Ho-doping content [45]. In addition, no additional optical transition, which could be associated with the SPS process, was observed in the absorption or luminescence spectra reported in Figures 5 and 7. Moreover, Capobianco et al. obtained a very similar evolution of Visible transition intensities for $Ho:Y_2O_3$ samples sintered under air [37]. As a result, the $Ho:Lu_2O_3$ ceramics elaborated in this study present very similar optical properties compared to ceramics sintered by other techniques, as well as those of single crystals. Nanopowder chemical synthesis coupled with SPS sintering technology appears to be a very promising elaboration process to obtain laser ceramics. However, the optical quality should be still improved, i.e., the residual porosity seen in Figure 4 should be totally removed. Also, the samples are directly in contact with the graphite die during sintering and often present a grey/dark color after sintering. This was assigned by many studies to a reduction in the material linked to carbon contamination, and it can be easily removed by a reoxidation treatment under air at a moderate temperature (i.e., 900–120 °C). Nevertheless, the time required for such a treatment evolves as a function of e^2, where e is the thickness of samples, because the rate of the reoxidation process is controlled by solid-state diffusion. As a result, a very long time is required for the reoxidation of samples with centimetric thickness, constituting a strong limitation of the efficiency of the whole process. Further investigations will be carried out on these two points to improve the optical quality of the ceramics and the scalability of the process.

Table 2. Spectroscopic characteristics of $Lu_{2-x}Ho_xO_3$ samples as a function of process parameters.

Reference	Composition	Material	Sintering	G_{mean} (μm)	T% at 2 μm	λ_{abs} (nm)	σ_{abs} (cm^2)	λ_{em} (nm)	σ_{em} (cm^2)	τ_{rad} (ms)
This work	1%Ho:Lu_2O_3	Ceramic	SPS + air annealing	<1 μm	80%	1927	1×10^{-20}	2030		11.5
Kim et al. [39]	2%Ho:Lu_2O_3	Ceramic	Post-HIP + air annealing	50 μm	>80%	1942	-	2140	-	10
Dong et al. [44]	1%Ho:Lu_2O_3	Single-crystal	-	-	>80%	1929	1.1×10^{-20}	2027	3.3×10^{-21}	9.5
Koopmann et al. [45]	0.3%Ho:Lu_2O_3	Single-crystal	-	-	>80%	1928	1.2×10^{-20}	2124	4.5×10^{-21}	10

5. Conclusions

$Lu_{2-x}Ho_xO_3$ transparent ceramics with a small grain size (<1 μm) were successfully manufactured by Spark Plasma Sintering of homemade nanopowders. The photoluminescence study performed in the Visible and IR domains has given similar results to those obtained for single crystals of similar compositions. Especially, cross-relaxation mechanisms have been identified and favor the 2 μm emission. Moreover, fluorescence lifetime measurements have highlighted concentration quenching, especially for high doping of 10% Ho. All of the obtained results indicate that the optical properties are very similar between $Lu_{2-x}Ho_xO_3$ transparent ceramics and single crystals. Thus, the distribution of Ho^{3+} ions should be similar in both materials, i.e., no clustering of Ho^{3+} ions in ceramics should take place, whatever the process parameters or the doping level. Finally, SPS appears to be a very promising method to produce transparent polycrystalline ceramics of the rare-earth oxide family. Among them, $Lu_{2-x}Ho_xO_3$ ceramics could substitute single crystals of similar composition and are very promising materials for high-energy solid-state lasers.

Author Contributions: Conceptualization, R.B. and L.V.; methodology, R.B. and L.V.; validation, R.B. and A.M.; formal analysis, L.V.; investigation, L.V. and S.G.-G.; data curation, L.V.; writing—original draft preparation, L.V.; writing—review and editing, R.B., V.J., G.D.-B., É.C. and A.M.; supervision, R.B. and A.M.; project administration, R.B.; funding acquisition, R.B. All authors have read and agreed to the published version of the manuscript.

Funding: This research was funded by the Region Nouvelle-Aquitaine (MISTRAL program n° 2017-1R50311 and HICEMIR program n° 2017-1R50311) and was also supported by an institutional grant from the National Research Agency under the investments for the future program with the reference ANR-10-LABX-0074-01 Sigma-LIM.

Institutional Review Board Statement: Not applicable.

Informed Consent Statement: Not applicable.

Data Availability Statement: The data presented in this study are available on request from the corresponding author. The data are not publicly available due to the data are part of an ongoing study.

Acknowledgments: The authors are grateful to Pierre Carles (IRCER, Limoges, France) for the TEM and STEM observations and Victor Anthony Garcia Rivera (COPL, University Laval) for the lifetime measurements performed in the frame of the International Associated Laboratory (LIA) LuMAQ.

Conflicts of Interest: The authors declare no conflicts of interest. The funders had no role in the design of the study; in the collection, analyses or interpretation of data; in the writing of the manuscript or in the decision to publish the results.

References

1. Ikesue, A.; Aung, Y.L. Ceramic laser materials. *Nat. Photon.* **2008**, *2*, 721–727. [CrossRef]
2. Sanghera, J.; Bayya, S.; Villalobos, G.; Kim, W.; Frantz, J.; Shaw, B.; Sadowski, B.; Miklos, R.; Baker, C.; Hunt, M.; et al. Transparent ceramics for high-energy laser systems. *Opt. Mater.* **2011**, *33*, 511–518. [CrossRef]
3. Kim, W.; Villalobos, G.; Baker, C.; Frantz, J.; Shaw, B.; Bayya, S.; Bowman, S.; Sadowski, B.; Hunt, M.; Rock, B.; et al. Overview of transparent optical ceramics for high-energy lasers at NRL. *Appl. Opt.* **2015**, *54*, F210–F221. [CrossRef]

4. Zhu, D.; Chen, Z.-Z.; Song, Y.-J.; Yuan, L.; Lin, Y.-Y.; Zhang, L.-N.; Cui, D.-F.; Bo, Y.; Peng, Q.-J.; Xu, Z.-Y.; et al. High Power (~10 kW) Yb:YAG Ceramic Slab Laser Operating at 1030 nm. *IEEE Photon. Technol. Lett.* **2023**, *35*, 789–792. [CrossRef]
5. Zhang, W.X.; Zhou, J.; Liu, W.B.; Li, J.; Wang, L.; Jiang, B.X.; Pan, Y.B.; Cheng, X.J.; Xu, J.Q. Fabrication, properties and laser per-formance of Ho:YAG transparent ceramic. *J. Alloys Compd.* **2010**, *506*, 745–748. [CrossRef]
6. Zhang, W.X.; Pan, Y.B.; Zhou, J.; Liu, W.B.; Li, J.; Jiang, B.X.; Cheng, X.J.; Xu, J.Q. Diode-Pumped Tm:YAG Ceramic Laser. *J. Am. Ceram. Soc.* **2009**, *92*, 2434–2437. [CrossRef]
7. Scholle, K.; Lamrini, S.; Koopmann, P.; Fuhrberg, P. 2 μm Laser Sources and Their Possible Applications. In *Frontiers in Guided Wave Optics and Optoelectronics*; Pal, B., Ed.; IntechOpen: London, UK, 2010.
8. Yang, H.; Zhang, J.; Qin, X.P.; Luo, D.W.; Ma, J.; Tang, D.Y.; Zhang, Q.T. Fabrication and Properties of High Quality Transparent Ho:YAG Ceramics. *Solid State Phenom.* **2012**, *185*, 51–54. [CrossRef]
9. Payne, S.A.; Chase, L.L.; Smith, L.K.; Kway, W.L.; Krupke, W.F. Infrared cross section measurements for crystals doped with Er^{3+}, Tm^{3+}, and Ho^{3+}. *IEEE J. Quantum Electron.* **1992**, *28*, 2619–2630. [CrossRef]
10. Wang, Y.; Lan, R.; Mateos, X.; Li, J.; Hu, C.; Li, C.; Suomalainen, S.; Härkönen, A.; Guina, M.; Petrov, V.; et al. Broadly tunable mode-locked Ho:YAG ceramic laser around 21 μm. *Opt. Express* **2016**, *24*, 18003–18012. [CrossRef] [PubMed]
11. Chen, H.; Shen, D.; Zhang, J.; Yang, H.; Tang, D.; Zhao, T.; Yang, X. In-band pumped highly efficient Ho:YAG ceramic laser with 21 W output power at 2097 nm. *Opt. Lett.* **2011**, *36*, 1575–1577. [CrossRef]
12. Cheng, X.; Xu, J.; Wang, M.; Jiang, B.; Zhang, W.; Pan, Y. Ho:YAG ceramic laser pumped by Tm:YLF lasers at room temperature. *Laser Phys. Lett.* **2010**, *7*, 351–354. [CrossRef]
13. Antipov, O.L.; Novikov, A.A.; Zakharov, N.G.; Zinoviev, A.P. Optical properties and efficient laser oscillation at 2066 nm of novel Tm:Lu_2O_3 ceramics. *Opt. Mater. Express* **2012**, *2*, 183–189. [CrossRef]
14. Ikesue, A.; Aung, Y.L.; Lupei, V. *Ceramic Lasers*; Cambridge University Press: Cambridge, UK, 2013; p. 165.
15. Salek, G.; Devoti, A.; Garcia, A.; Gaudon, M.; Jubera, V.; Demourgues, A. Tuning the composition of rare earth sesquioxides $Gd_{2-x}La_xO_3$:Eu^{3+} to control phase transitions at a high temperature to design new highly sensitive luminescence-based thermal sensors. *R. Soc. Chem.* **2016**, *6*, 298–306. [CrossRef]
16. Permin, D.; Kurashkin, S.; Novikova, A.; Savikin, A.; Gavrishchuk, E.; Balabanov, S.; Khamaletdinova, N. Synthesis and luminescence properties of Yb-doped Y_2O_3, Sc_2O_3 and Lu_2O_3 solid solutions nanopowders. *Opt. Mater.* **2018**, *77*, 240–245. [CrossRef]
17. Tang, F.; Huang, J.; Guo, W.; Wang, W.; Fei, B.; Cao, Y. Photoluminescence and laser behavior of Yb:YAG ceramic. *Opt. Mater.* **2012**, *34*, 757–760. [CrossRef]
18. Unal, F.; Kaya, F. Kazmanli, KEffects of dopant rate and calcination parameters on photoluminescence emission of Y_2O_3:Eu^{3+} phosphors: A statistical approach. *Ceram. Int.* **2019**, *45*, 17818–17825. [CrossRef]
19. Malinowski, M.; Kaczkan, M.; Wnuk, A.; Szuflińska, M. Emission from the high lying excited states of Ho^{3+} ions in YAP and YAG crystals. *J. Lumin.* **2004**, *106*, 269–279. [CrossRef]
20. Liu, Z.; Ikesue, A.; Li, J. Research progress and prospects of rare-earth doped sesquioxide laser ceramics. *J. Eur. Ceram. Soc.* **2021**, *41*, 3895–3910. [CrossRef]
21. Fu, Z.; Wu, N.; Long, H.; Hou, Z. Fabrication of fine-grained and high thermal properties Y_2O_3 transparent ceramics without sintering aids. *Opt. Mater.* **2024**, *147*, 114618. [CrossRef]
22. Liu, Z.; Feng, Y.; Toci, G.; Pirri, A.; Patrizi, B.; Vannini, M.; Li, J. Influence of annealing on microstructures and properties of Yb:Lu_2O_3 transparent ceramics. *J. Am. Ceram. Soc.* **2023**, *107*, 1974–1984. [CrossRef]
23. Liu, Z.; Toci, G.; Pirri, A.; Patrizi, B.; Feng, Y.; Hu, D.; Chen, H.; Hreniak, D.; Vannini, M.; Li, J. Fabrication and characterizations of Tm:Lu_2O_3 transparent ceramics for 2 μm laser applications. *Opt. Mater.* **2022**, *131*, 112705. [CrossRef]
24. Maksimov, R.; Shitov, V.; Osipov, V.; Samatov, O.; Vakalov, D.; Malyavin, F.; Basyrova, L.; Loiko, P.; Camy, P. Fabrication, microstructure and mid-infrared luminescence of Er:$(Sc_xY_{1-x})_2O_3$ transparent ceramics. *Opt. Mater.* **2023**, *137*, 113542. [CrossRef]
25. Chaim, R.; Levin, M.; Shlayer, A.; Estournes, C. Sintering and densification of nanocrystalline ceramic oxide powders: A review. *Adv. Appl. Ceram.* **2008**, *107*, 159–169. [CrossRef]
26. Morita, K.; Kim, B.-N.; Yoshida, H.; Hiraga, K.; Sakka, Y. Distribution of carbon contamination in oxide ceramics occurring during spark-plasma-sintering (SPS) processing: II Effect of SPS and loading temperatures. *J. Eur. Ceram. Soc.* **2018**, *38*, 2596–2604. [CrossRef]
27. Alombert-Goget, G.; Guyot, Y.; Guzik, M.; Boulon, G.; Ito, A.; Goto, T.; Yoshikawa, A.; Kikuchi, M. Nd^{3+}-doped Lu_2O_3 transparent sesquioxide ceramics elaborated by the Spark Plasma Sintering (SPS) method. Part 1: Structural, thermal conductivity and spectroscopic characterization. *Opt. Mater.* **2015**, *41*, 3–11. [CrossRef]
28. Boulesteix, R.; Epherre, R.; Noyau, S.; Vandenhende, M.; Maître, A.; Sallé, C.; Alombert-Goget, G.; Guyot, Y.; Brenier, A. Highly transparent Nd:Lu_2O_3 ceramics obtained by coupling slip-casting and spark plasma sintering. *Scr. Mater.* **2014**, *75*, 54–57. [CrossRef]
29. Lupei, V.; Lupei, A.; Boulon, G.; Jouini, A.; Ikesue, A. Assessment of the distribution of the Yb^{3+} ions in Sc_2O_3 ceramics from cooperative absorption and emission. *J. Alloys Compd.* **2008**, *451*, 179–181. [CrossRef]
30. Liu, Z.; Toci, G.; Pirri, A.; Patrizi, B.; Li, J.; Hu, Z.; Wei, J.; Pan, H.; Xie, T.; Vannini, M.; et al. Fabrication and laser operation of Yb:Lu2O3 transparent ceramics from co-precipitated nano-powders. *J. Am. Ceram. Soc.* **2019**, *102*, 7491–7499. [CrossRef]

31. Thoř, T.; Rubešová, K.; Jakeš, V.; Mikolášová, D.; Cajzl, J.; Havlíček, J.; Nádherný, L.; Průša, F.; Kučerková, R.; Nikl, M. Dense ceramics of lanthanide-doped Lu_2O_3 prepared by spark plasma sintering. *J. Eur. Ceram. Soc.* **2021**, *41*, 741–751. [CrossRef]
32. Viers, L.; Delaunay, F.; Boulesteix, R.; Vandenhende, M.; Antou, G.; Maître, A. Study of densification mechanisms during Spark Plasma Sintering of co-precipitated Ho:Lu_2O_3 nanopowders: Application to transparent ceramics for lasers. *J. Eur. Ceram. Soc.* **2021**, *41*, 7199–7207. [CrossRef]
33. Kaminskii, A.A.; Akchurin, M.S.; Becker, P.; Ueda, K.; Bohatý, L.; Shirakawa, A.; Tokurakawa, M.; Takaichi, K.; Yagi, H.; Dong, J. Mechanical and optical properties of Lu_2O_3 host-ceramics for Ln^{3+} lasants. *Laser Phys. Lett.* **2008**, *5*, 300–303. [CrossRef]
34. Boulesteix, R.; Maître, A.; Baumard, J.-F.; Rabinovitch, Y.; Reynaud, F. Reynaud, Light scattering by pores in transparent Nd:YAG ceramics for lasers: Correlations between microstructure and optical properties. *Opt. Express* **2010**, *18*, 14992–15002. [CrossRef]
35. Brown, D.C.; Envid, V.; Zembek, J. Ho:YAG absorption cross sections from 1700 to 2200 nm at 83, 175, and 295 K. *Appl. Opt.* **2012**, *51*, 8147–8158. [CrossRef]
36. Silver, J.; Barrett, E.; Marsh, P.J.; Withnall, R. Yttrium Oxide Upconverting Phosphors. 5. Upconversion Luminescent Emission from Holmium-Doped Yttrium Oxide under 632.8 nm Light Excitation. *J. Phys. Chem. B* **2003**, *107*, 9236–9242. [CrossRef]
37. Capobianco, J.A.; Boyer, J.C.; Vetrone, F.; Speghini, A.; Bettinelli, M. Optical Spectroscopy and Upconversion Studies of Ho3+-Doped Bulk and Nanocrystalline Y_2O_3. *Chem. Mater.* **2002**, *14*, 2915–2921. [CrossRef]
38. Auzel, F.; Baldacchini, G.; Laversenne, L.; Boulon, G. Radiation trapping and self-quenching analysis in Yb^{3+}, Er^{3+}, and Ho^{3+} doped Y_2O_3. *Opt. Mater.* **2003**, *24*, 103–109. [CrossRef]
39. Kim, W.; Baker, C.; Bowman, S.; Florea, C.; Villalobos, G.; Shaw, B.; Sadowski, B.; Hunt, M.; Aggarwal, I.; Sanghera, J. Laser oscillation from Ho^{3+} doped Lu_2O_3 ceramics. *Opt. Mater. Express* **2013**, *3*, 913–919. [CrossRef]
40. Kijko, V.; Maksimov, R.N.; Shitov, V.; Demakov, S. Yurovskikh, ASintering of transparent Yb-doped Lu_2O_3 ceramics using nanopowder produced by laser ablation method. *J. Alloys Compd.* **2015**, *643*, 207–211. [CrossRef]
41. Toci, G.; Vannini, M.; Ciofini, M.; Lapucci, A.; Pirri, A.; Ito, A.; Goto, T.; Yoshikawa, A.; Ikesue, A.; Alombert-Goget, G.; et al. Nd^{3+}-doped Lu_2O_3 transparent sesquioxide ceramics elaborated by the Spark Plasma Sintering (SPS) method. Part 2: First laser output results and comparison with Nd^{3+}-doped Lu_2O_3 and Nd^{3+}-Y_2O_3 ceramics elaborated by a conventional method. *Opt. Mater.* **2014**, *41*, 12–16. [CrossRef]
42. Xu, C.; Yang, C.; Zhang, H.; Duan, Y.; Zhu, H.; Tang, D.; Huang, H.; Zhang, J. Efficient laser operation based on transparent Nd:Lu_2O_3 ceramic fabricated by Spark Plasma Sintering. *Opt. Express* **2016**, *24*, 20571–20579. [CrossRef] [PubMed]
43. Boulon, G.; Epicier, T.; Zhao, W.; Guzik, M.; Pan, Y.; Jiang, B. *Is There Segregation of Rare Earth Ions in Garnet Optical Ceramics?* Bartolo, B.D., Collins, J., Baldassare, D.B., John, C., Eds.; Nano-Optics for enhancing light-matter interactions on a molecular scale. NATO Science for Space and Security Series B: Physics and Biophysics; Springer: Dordrecht, The Netherlands, 2013. [CrossRef]
44. Dong, J.; Wang, W.; Xue, Y.; Hou, W.; Cao, X.; Xu, X.; Wu, F.; Luo, P.; Wang, Q.; Li, D.; et al. Crystal growth and spectroscopic analysis of Ho:Lu_2O_3 crystal for mid-infrared emission. *J. Lumin.* **2022**, *251*, 119192. [CrossRef]
45. Koopmann, P.; Lamrini, S.; Scholle, K.; Schäfer, M.; Fuhrberg, P.; Huber, G. Multi-watt laser operation and laser parameters of Ho-doped Lu_2O_3 at 2.12 μm. *Opt. Mater. Express* **2011**, *1*, 1447–1456. [CrossRef]

ceramics

MDPI

Article

Glass Composition for Coating and Bonding of Polycrystalline Spinel Ceramic Substrates

Jacob Hormadaly

Department of Chemistry, Ben-Gurion University, Hashalom St. #1, P.O. Box 653, Beer-Sheva 84105, Israel; hormadj@bgu.ac.il

Abstract: Design considerations of the lead-based glass composition was broadened beyond the two known criteria of matched index of refraction and thermal coefficient of expansion to include previous studies of thick film materials. Five criteria for the glass-design composition were used: matched index of refraction and thermal coefficient of expansion, components (MgO and Al_2O_3) to slow down dissolution of spinel ($MgAl_2O_4$) into the glass, non-crystallizing glass in a broad temperature range and glass with good chemical durability. Synthesis and characterization of glass, glass paste preparation and its application to spinel substrates to form coating and bonding and optical characterizations in the UV, VIS and IR of coated, uncoated, and bonded spinel substrates of two commercial sources are described. Enhancement of transmittance exceeding the theoretical value of polished spinel was found for the first time when glass coating was applied to a ground face of semi-polished spinel.

Keywords: glass coating and bonding; spinel; optical properties

1. Introduction

The cost of production of ceramic parts in current technologies, and specifically, in single crystal sapphire (Al_2O_3) and ceramics comprising spinel structures ($MgAl_2O_4$), is high due to an expansive processing and polishing procedure. In ceramic polishing, the expansive machining or polishing of the surfaces is unavoidable in most cases in which the end product is required to transmit light. The difficulties involved in polishing ceramics and particularly spinel structures can result in a large fraction of the total cost being associated with surface finishing and polishing. To reduce these high polishing and finishing costs, a refractive index matched glass coating is applied to the hard inorganic substrates to render them transparent. The fired glass coating can be polished if required using conventional polishing methods which are very cheap. The feasibility of coating a hard inorganic material such as sapphire with a lead-based glass frit was demonstrated on small tiles [1] and on large area windows which were coated and bonded with glass [2]. The selection criteria for the glass compositions used for coating in the cited above works were index of refraction and probably linear coefficient of expansion matching to the ceramic substrate. Are these criteria sufficient for lead-based glass for coating sapphire and spinel? The answer is no if we consider studies in thick film technology and materials. High lead-based glasses which are used in thick film materials are very reactive and interact with the ceramic substrates (96% Al_2O_3, 99% Al_2O_3, sapphire) at moderate temperatures. These glasses dissolve substantial amount of the substrate [3,4] during firing at Birox® profile (fast ramp to 850 °C dwell of 10 min at 850 °C, fast cool down to room temperature, the whole profile is 30 to 60 min). Dissolution of the substrates into the glass changes the properties of the glass.

Spinel ceramic ($MgAl_2O_4$) is a durable, isotropic material with outstanding optical and mechanical properties and has been the subject of intensive research efforts in the last 70 years [5–9]. Some of its diverse applications were reported in reference [5] and other applications are collected in references [10–19].

The size of the substrates synthesized was scaled up to large windows [11,12] that can be used for transparent armor as a strike-face layer. Large substrates have coarse-grained structure; evaluation [20] "showed that, under the same loading conditions, the flexure strength of coarse-grained spinel was approximately 75 MPa while the strength of a fine-grained (i.e., <25 μm), sintered spinel was close to 200 MPa". High-strength substrates are still limited in size, and a suitable bonding material is required to make large windows.

Low-fire high-barium leadless borosilicate glass frit, Pemco-626, was used to coat unpolished and polished spinel substrates [21]. The glass has TCE similar to spinel, measured TCE $7.3 \times 10^{-6}/°C$, calculated [22] TCE $7.6 \times 10^{-6}/°C$, but the index of refraction is not matched to spinel. The glass adhered very well to the spinel, and coating improved the transparency of unpolished spinel, but transparency is still low. Recently, a lead borosilicate glass composition was suggested for joining (edge joining) spinel tiles [23]. The glass matched the index of refraction, and TCE to spinel and joining was performed at 760 °C.

This study is based on a patent application [24]; it deals with modified lead silicate composition (melted in 2015), which contains Al_2O_3, MgO, ZnO and TiO_2 in addition to PbO and SiO_2. Five criteria were used to design the composition: index of refraction and TCE matched to spinel, glass ingredients (Al_2O_3, MgO) to slow down dissolution of spinel into the glass, non-crystallizing glass composition at a broad temperature range, and a durable glass composition. The paper also details synthesis of glass, glass paste, measurements of optical properties of coated and uncoated spinel substrates, edge, and face-to-face bonding of spinel tiles. In addition, to lead-based glass, the paper also details the synthesis and some properties of leadless-glass compositions based on an La_2O_3-MgO-ZnO-Al_2O_3-B_2O_3-SiO_2 system.

2. Experimental

Analytical reagent grade SiO_2, PbO, MgO, ZnO, TiO_2, Al $(OH)_3$ and Hammond Lead glass B15 powder (composition in wt.% SiO_2-34; PbO-63:Al_2O_3-2: TiO_2-1) were used as batch materials. Batch materials were mixed in a polyethylene container then placed into Pt crucible and heated in a box furnace at 1200–1400 °C to melt and homogenize for 0.5 to 3 h. Glass melt was poured into water to obtain glass frit. Glass frit dried in the oven at 150 °C, and then it was ball milled with alumina grinding media (200 g raw frit and 200 mL isopropanol) for 24 h. After separation of the alumina grinding media, the slip was passed through 325 and 400 mesh stainless screens and let evaporate in the hood, and then they were dried in an oven at 150 °C for several hours. The batch size was about 500 g and four batches were melted.

Before fritting, the melt was cast into pre-heated box-shaped stainless-steel modular molds to form glass bars of $5 \times 0.8 \times 0.8$ cm^3. The bars were annealed for 0.5 h in an electric box furnace. Annealing temperature (T_{an}) was estimated according to the formula $T_{an} = T_g + 10$ °C. The glass bars were cut and polished to produce parallel faces. Linear coefficients of expansion were in the temperature range of 25–300 °C ($TCE_{25–300}$); the glass transition temperatures (T_g) and the dilatometric softening points (T_d) were measured with an Orton dilatometer, model 1000D, at a heating rate of 3 °C/min. The measurement accuracy of the linear coefficient of expansion was ±10%. Part of the expansion bar (after annealing) was cut to provide a small piece, which was polished and used to measure the index of refraction at the sodium D line. Index of refraction was measured with Atago company, Japan refractometers models 4T and DR-A1.

Glass compositions of the system La_2O_3-MgO-ZnO-Al_2O_3-B_2O_3-SiO_2 were synthesized from analytical reagent grade H_3BO_3, SiO_2, Al $(OH)_3$, ZnO, MgO and La_2O_3. MgO and La_2O_3 were first heated at 1000 °C for one hour to decompose hydroxides and carbonates which form in storage. Batch ingredients were melted in Pt crucible at the temperature range of 1350 to 1450 °C and kept at peak temperature for 1 to 3 h, and then glass bars were cast as described above and annealed at 660 °C. Batch size was about 50 g. Thermal expansion and the index of refraction were measured as described above. Glass powder was made by crushing the glass bar to a coarse powder, and then the coarse powder was milled in a Spex sample prep. model 8000M (Metuchen, NJ, USA) for 60 min.

The glass paste consists of inorganic particulate (glass powder) and organic materials. The organic materials used consisted of a mixture of diethylene glycol dibutyl ether, terpineol, and ethyl cellulose. The ingredients of the paste are weighed together in a container and vigorously mixed to form a uniform blend; then, the blend is passed through dispersing equipment, such as a Muller, to achieve a good dispersion of particles. Preparation of the glass paste can also be done by grinding the paste ingredients (glass powder 70%, solvents (terpineol and diethylene glycol dibutyl ether) 20% and medium 10% (medium is a solution of ethyl cellulose10% in terpineol 45% and diethylene glycol dibutyl ether 45%)) in an agate mortar and pestle. For a larger quantity of paste a three-roll mill can be used. In this study the paste was applied to the spinel substrates manually by spreading it with a spatula; for large area substrates, a screen printer maybe used.

The coating is a one-stage process; paste is applied (thickness of coating is controlled using the weight of the paste, and the thickness of the fired glaze is about 50 microns) to the surface or edge of a substrate (two types of commercial spinel substrates were used: one made by Ceramtech and the second by Rafael companies), and paste is spread to cover the whole face or the edge, and then the coated substrate is left at room temperature so the paste can level. After leveling, the coated substrate is heated in an oven at 150 °C for 30 min to evaporate a large fraction of the solvents to obtain dry paste (the dry paste still contains the ethyl cellulose and a small part of the solvents). After drying, the coated substrate is fired in profile I (Figure 1). Profile I has two slow-heating segments which allow evaporation, pyrolysis, and burning of the organic materials, and then fast ramping at 20 °C/min to 1100 °C, dwelling of 24 min at 1100 °C then cooling the segment at 20 °C/min to 560 °C, and then, dwelling for 1 h to anneal the glass coating and then cooling it down at 20 °C/min to room temperature. After profile I, a glazed surface or edge is obtained. The bonding or the joining of two faces or edges is a two-stage process; the faces or edges of two substrates are processed as described above using profile I to obtain two glazed faces or edges. Glazed faces or edges are placed one on top of the other, glazed faces or edges are in contact and fired in profile II (Figure 2). Profile II has only fast ramping at 20 °C/min to 1100 °C, dwelling of 24 min at 1100 °C, then cooling of the segment at 20 °C/min to 560 °C, and then dwelling for 1 h to anneal the glass and then cooling it down at 20 °C/min to room temperature. After profile II, two bonded substrates at faces or edges are obtained.

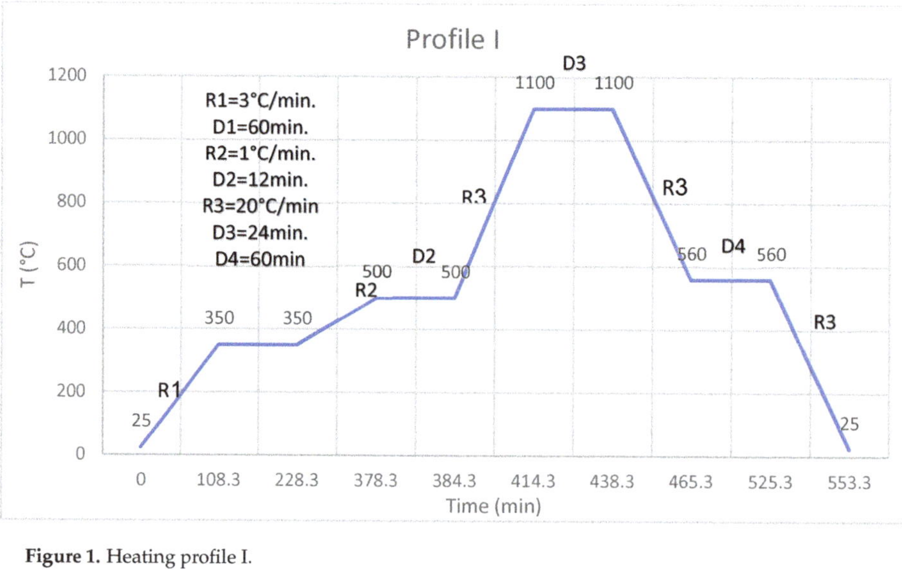

Figure 1. Heating profile I.

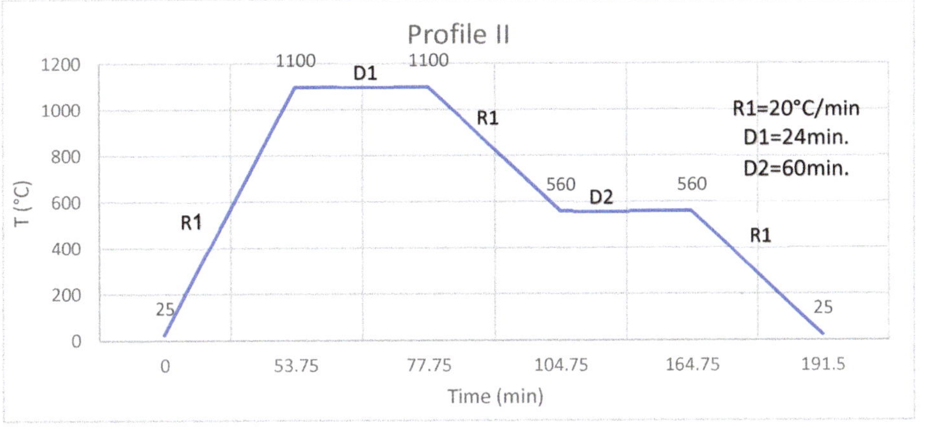

Figure 2. Heating profile II.

Surface area was measured using the BET method using Quantachrome Nova Touch LX3 (FL, USA). Surface area of ball-milled glass powder was 1.63 m^2/g.

Spectra in the UV-VIS and IR ranges were measured with Jasco V-530 and Thermo Nicolet 6700, respectively.

3. Results and Discussion

Table 1 presents the composition, in mole% and wt.%, of the glass of this study. According to the design criteria, the composition contains about 6 mole% of alumina and magnesia to slow down dissolution of spinel into the glass. The references cited [3,4] conducted a comprehensive study on the lead-based-glass compositions and analyzed the alumina content dissolved into the glass and correlated it with the glass composition. As the alumina content in the lead silicate increased, the glass dissolved less alumina from the substrate. The following citation is from the conclusions section of reference [4]: "These results suggest that the glass frit component of a resistive paste should contain an adequate amount of alumina in order to prevent further dissolution, to get a constant composition in the film, and so, uniform electrical properties of thick-film resistors". This supports the design criteria. Spinel is a source of alumina and magnesia. Inclusion of alumina and magnesia in the glass will slow or prevent dissolution of spinel in the glass. This assumption also follows from Le Chatelier's principle. Inclusion of alumina and magnesia also raises the viscosity of the glass, and higher glass viscosity slows down the kinetics of spinel dissolution into the glass.

Table 1. Glass composition.

Ingredient	Mole Percent	Weight Percent (wt.%)
PbO	25.83	54.89
MgO	5.98	2.3
ZnO	5.98	4.63
Al_2O_3	5.98	5.81
SiO_2	55.08	31.51
TiO_2	1.13	0.86

Lead oxide controls the viscosity and increases the refractive index, and titania increases the index of refraction and the glass durability. Zinc oxide was added to lower thermal expansion and to add another modifier to suppress crystallization.

The lead-based glass of Table 1 is unique: it does not crystallize on repeated heating at 1100 °C; it has low viscosity at high temperatures; it has TCE and N_d matched to spinel; and very good chemical durability.

Table 2 collects the glass transition T_g, dilatometer softening point T_d, linear coefficient of expansion in the temperature range of 25–300 °C, $TCE_{25–300}$, and the refractive index N_d of four glass batches melted. The calculated [22] TCE is 5.9×10^{-6}. The calculated TCE is in good agreement with the measured values. All four batches have consistent values of T_g, T_d, and N_d.

Table 2. Properties of glass compositions of batches 1, 2, 3, and 4.

Glass Code	Tg (°C)	Td (°C)	CTE $_{100–300°C}$ (10^{-6}/°C)	N_d
Batch 1	568	605	7.02	1.727
Batch 2	553	602	6.1	1.727
Batch 3	565	606	5.89	1.719
Batch 4	564	610	6	1.728

Glass compositions based on the La_2O_3-MgO-ZnO-Al_2O_3-B_2O_3-SiO_2 system are shown in Table 3. Properties of the glasses of Table 3 are collected in Table 4. The lanthanum-based glasses prepared have high T_g, TCE compatible with spinel, and N_d like spinel. However, the lanthanum-based system crystallizes when heated above 850 °C and is not suitable for coating spinel.

Table 3. Glass compositions of the system La_2O_3-MgO-ZnO-Al_2O_3-B_2O_3-SiO_2.

	Mole Percent		
Ingredient	**I**	**II**	**III**
La_2O_3	17.75	22.00	24.00
MgO	6.00	6.00	6.00
ZnO	12.00	12.00	12.00
Al_2O_3	12.25	14.00	12.00
B_2O_3	15.00	11.00	15.00
SiO_2	37.00	35.00	31.00

Table 4. Properties of glasses of the system La_2O_3-MgO-ZnO-Al_2O_3-B_2O_3-SiO_2.

Sample	T_g °C	T_d °C	$TCE_{100–300°C}$ (10^{-6})	N_d
I	688	726	7.2	1.717
II	-	-	7.6	1.748
III	702	745	8.2	1.774

Figures 1 and 2 were detailed in the experimental section. The dwell of 24 min at the high temperature of 1100 °C, well above the T_g, allows the release of gas bubbles from the glass coating because of the low viscosity of the glass at this temperature. Release of gas bubbles improves the optical quality of the glass coating.

Figure 3 shows the transmittances of polished substrates of Rafael company in the UV-VIS range: -33a is single substrate (two reflections), -33b is two substrates stacked together with cellotape on the edges (four reflections), and -33c is three substrates stacked together with cellotape on the edges (six reflections).The single substrate (-33a) is transparent in the UV down to 200 nm, and its transmittance in the VIS is high, almost like the theoretical value 86.8.

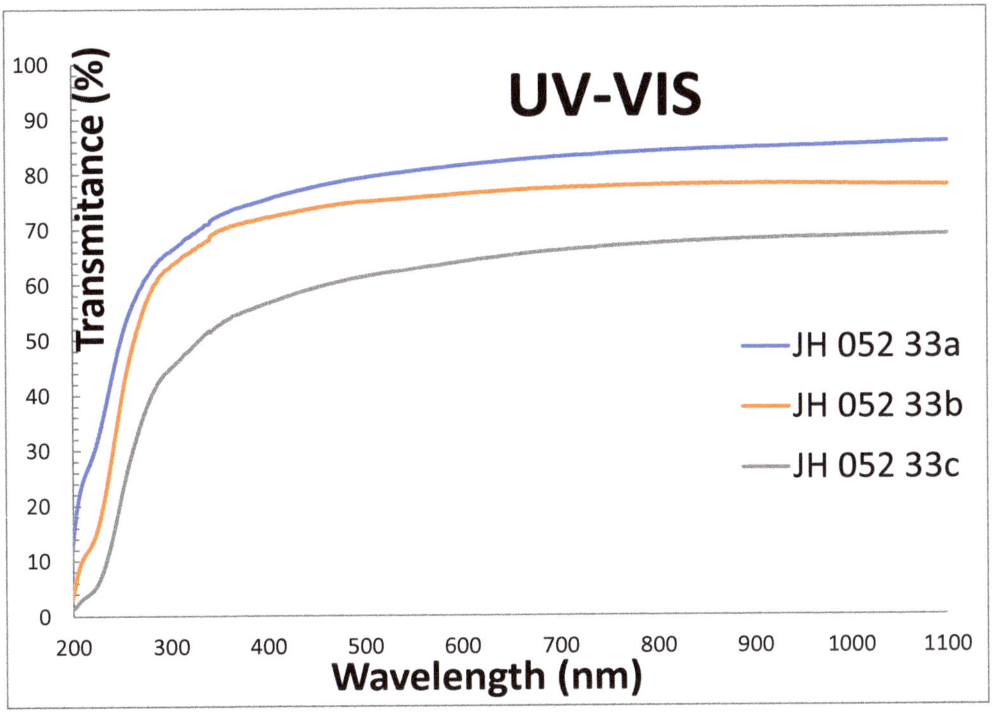

Figure 3. UV-VIS transmittances of polished spinel substrates (Rafael).

Figure 4 shows the IR transmittances of the substrates of Figure 3. The figure shows a small absorption band around 3380 cm^{-1}, which is probably the absorption of OH groups in the ceramic and an indication of a processing stage which involved water (milling in water) during the preparation of the spinel.

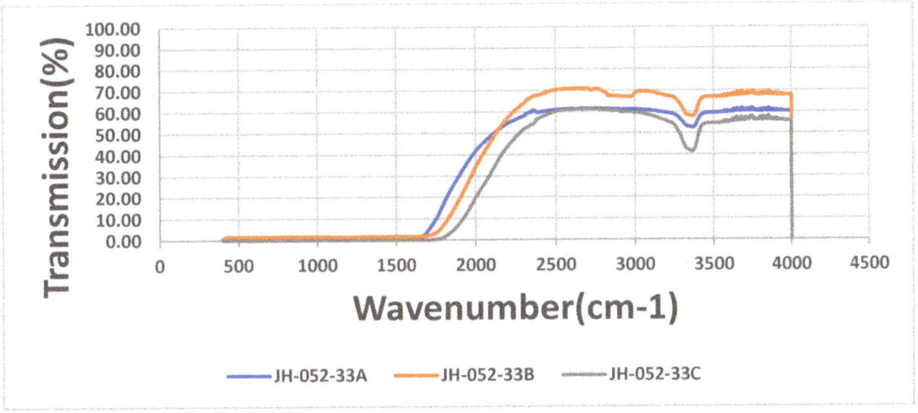

Figure 4. IR transmittances of polished spinel substrates (Rafael).

Figure 5 shows the transmittance in the UV-VIS range of coated semi-polished spinel of Rafael. Figure 5 shows high transmittance in the VIS range that is larger than that of a polished substrate. This result is unexpected because a glass coating on a polished spinel surface usually decreases the transmittance; see figure "transmittances of polished spinel (JH-052-100A) and polished spinel coated with glass on one face (JH-052-100B1)". The surface of the spinel (ground) may be the source of enhanced transmission shown in Figure 5, but further studies are required to better understand this unusual behavior.

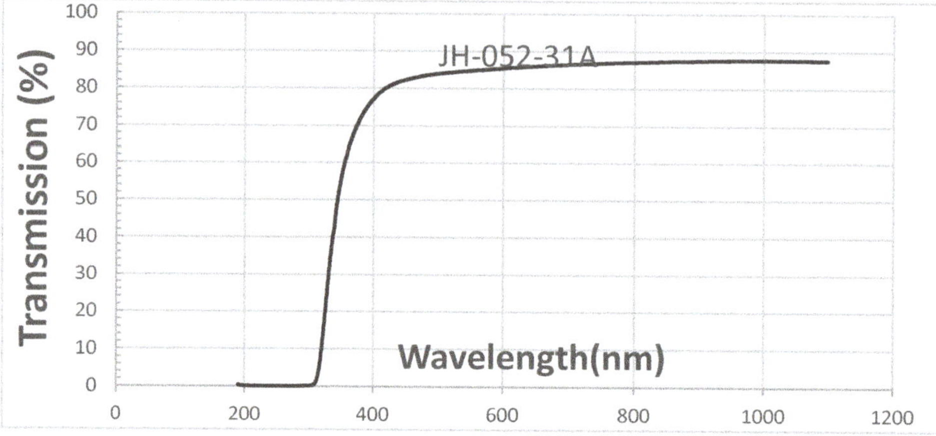

Figure 5. Transmittance of Rafael spinel semi-polished (one face polished, second face ground, coated with glass paste on ground face and fired).

The comparison of the transmittances of polished (-33a) and coated semi-polished (-31a) spinel substrates in the UV-VIS range is shown in Figure 6. Figure 6 shows that the uncoated polished spinel (-33a) is transparent in the UV almost down to 200 nm. The coated semi-polished spinel (-31a) has a total absorption of up to about 310 nm, and its transmittance in the VIS is larger than the theoretical value (86.8%). The absorption in the UV is due to an allowed $^1S_0$ to $^3P_1$ transition of Pb^{2+} ions in the glass. The transmittance of the semi-polished coated spinel (-31A) is larger than that of the polished spinel (-33A).

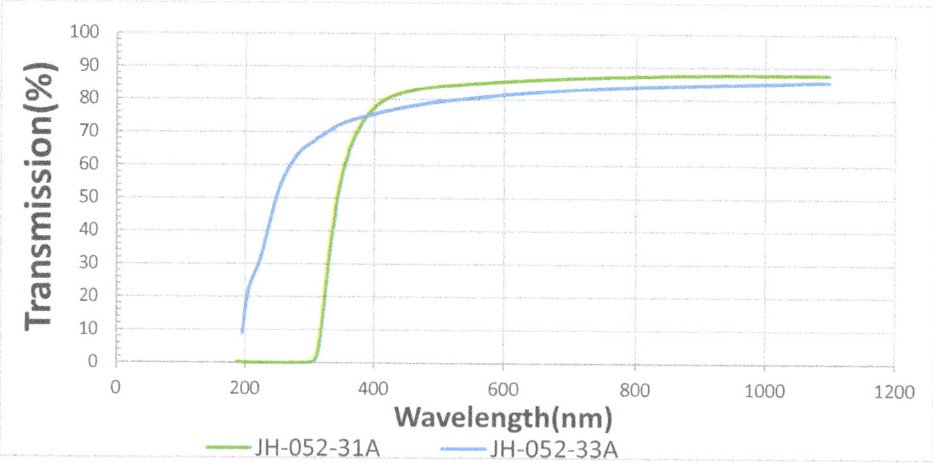

Figure 6. Comparison of transmittances of polished (JH-052-33A) and glass-coated spinel (JH-052-31A) in the UV-VIS semi-polished.

The comparison of the transmittances of polished (-33a) and coated semi-polished (-31a) spinel substrates in the IR range is shown in Figure 7. Figure 7 shows the IR transmittances of Figure 6's substrates; the glass-coated semi-polished spinel (-31a) has better transmittance than the polished (-33a) substrate and the glass coating did not add new absorption bands, and only below 2177 cm^{-1} did the polished substrate have slightly better transmittance.

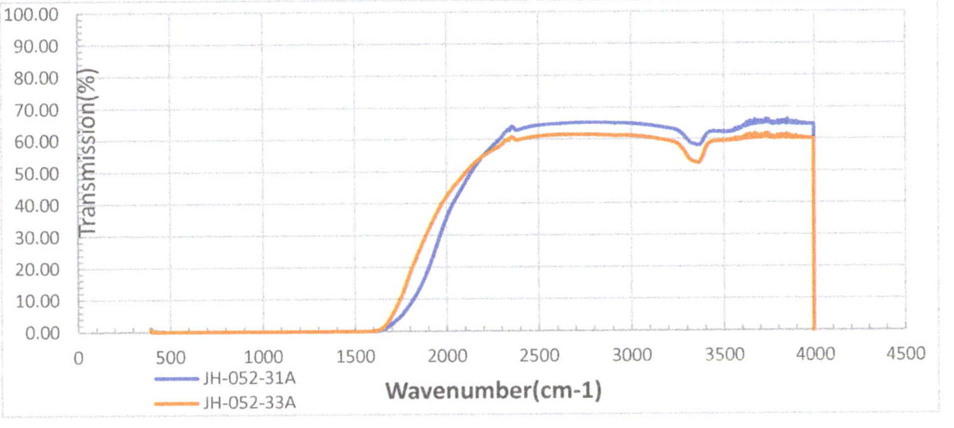

Figure 7. Comparison of transmittances of polished (JH-052-33A) and glass-coated semi-polished spinel (JH-052-31A) in the IR.

Figures 8 and 9 show the transmittances of polished Ceramtech substrates at the UV-VIS and IR ranges, respectively. In the figures, the transmittance of a single substrate (two reflections), two substrates stacked together with cellotape on the edges (four reflections), and three substrates stacked together with cellotape (six reflections) on the edges are shown to compare them to Figures 3 and 4 which show the substrates of Rafael. Ceramtech substrates (Figure 8) have absorption in the UV of up to 245 nm; Rafael's substrates have no absorption in the same wavelength range. The transmittances of Ceramtech's in the VIS are similar to those of Rafael's substrates. Figure 9 shows that Ceramtech spinel has large absorption bands in the IR at 3370 cm^{-1} and a shoulder at 3550 cm^{-1}. The absorption bands in the IR are the absorption of OH groups in the ceramic and an indication of a processing stage which involved water (milling in water) during the preparation of the spinel.

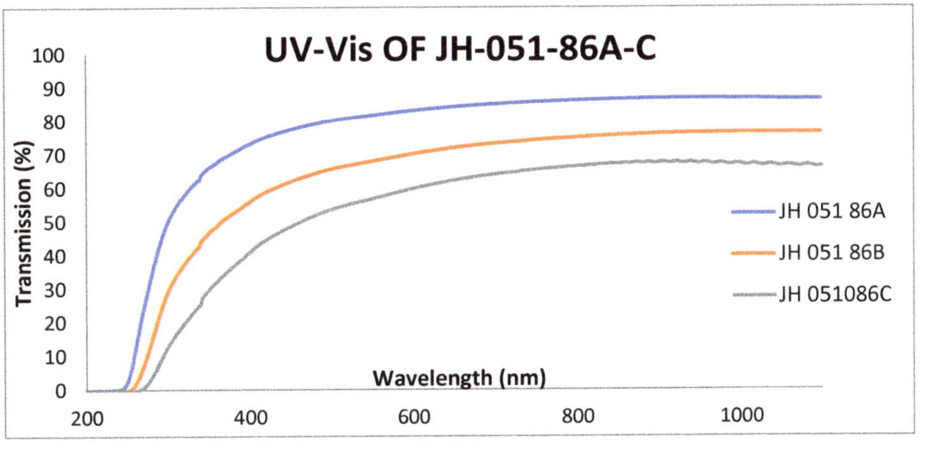

Figure 8. UV-Vis of polished substrates JH-051-86A-C (Ceramtech).

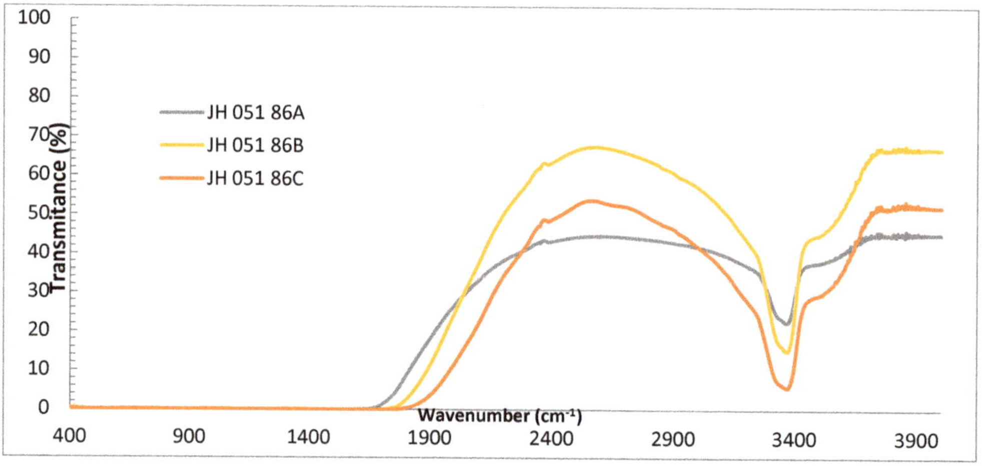

Figure 9. FTIR of polished substrates JH-051-86A-C (Ceramtech).

Figure 10 shows a comparison of transmittances of Rafael glass bonded (two polished substrates, face-to-face bonded with glass) with two polished spinel substrates stacked together with cellotape on the edges. The glass-bonded (-36) spinel has the expected absorption in the UV, but its transmittance at the VIS is high and similar to the transmittance of a single polished spinel. The transmittance of the stacked polished spinel is lower in the VIS because of four reflections; the glass-bonded spinel behaves like it has only two reflections.

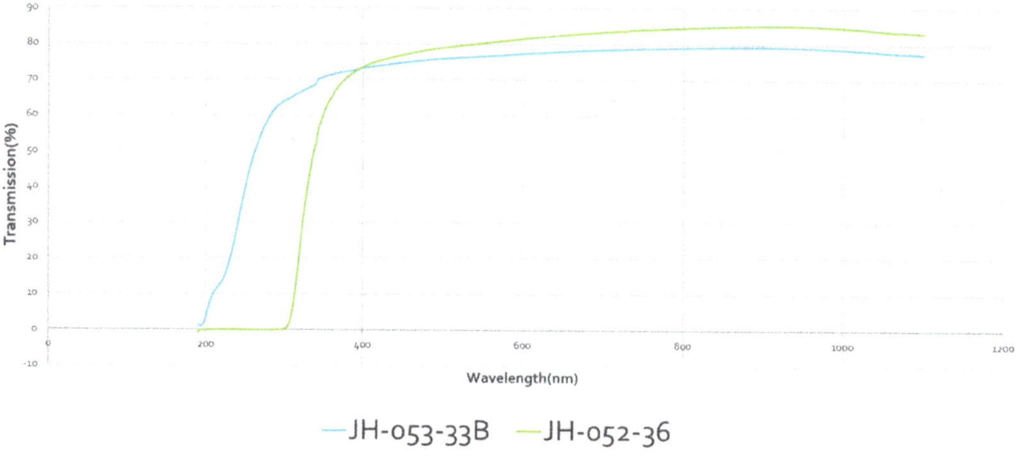

Figure 10. Comparison between two Rafael glass-bonded spinels (-36) and two stacked polished spinels (-33B).

Figure 11 shows the IR spectra of the substrates of Figure 10 (Rafael). The glass-bonded spinel (-36) has better transmittance in the IR, and the glass did not have any noticeable absorption bands.

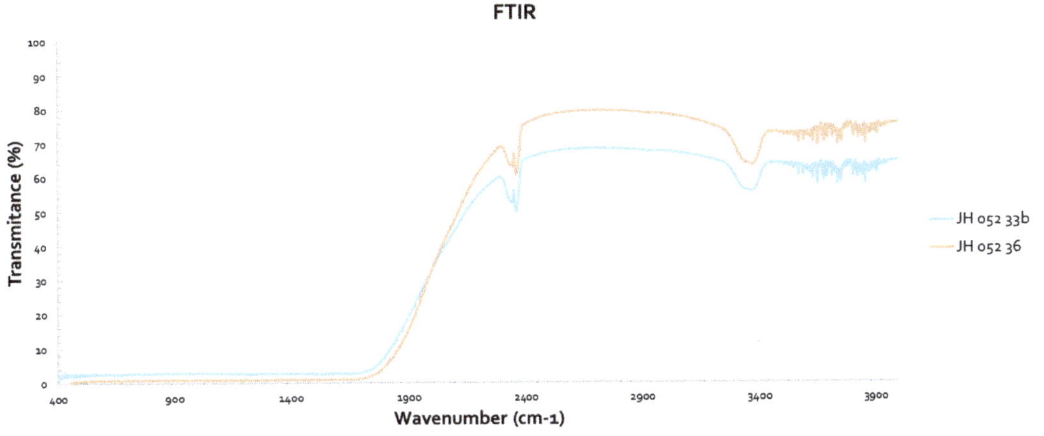

Figure 11. Comparison between two Rafael glass-bonded spinels (-36) and two polished spinels (-33B).

Figure 12 shows the UV-VIS transmittances of Rafael polished substrates:

-33A is single spinel, 0.8 cm thick; -33B is two spinel substrates taped together with cellotape on the edges, 1.6 cm thick; and -36 is two substrates face-to-face bonded with glass, 1.6 cm thick. The two-glass-bonded (-36) unit has VIS transmission similar to a single spinel, and almost identical. The cellotape-stacked-two-spinel unit has a lower transmission in the VIS because of the four reflections.

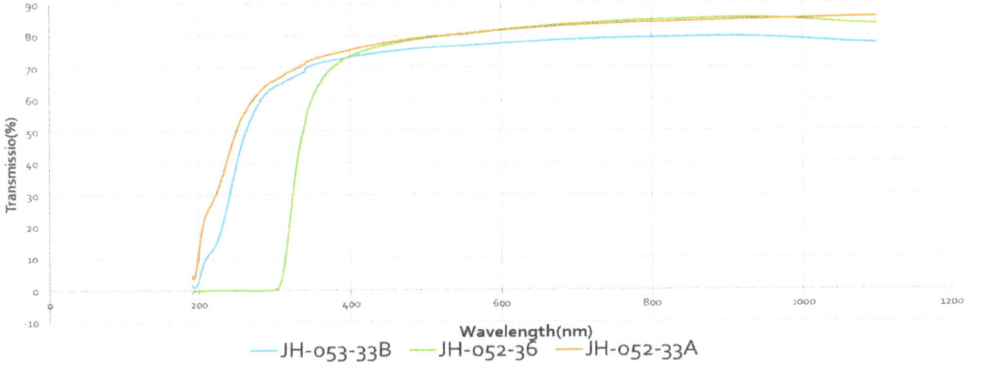

Figure 12. Comparison among polished single, two substrates taped together and two substrates bonded by a glass (Rafael).

JH-052-33 A: polished spinel 0.8 cm thick.

JH-052-33B: two polished spinels taped together at the edges 1.6 cm thick.

JH-052-36: two face to face glass bonded spinels 1.6 cm thick.

Figures 13 and 14 are pictures of edge-bonded spinels (Ceramtech) to demonstrate formation of a large area substrate from small substrates.

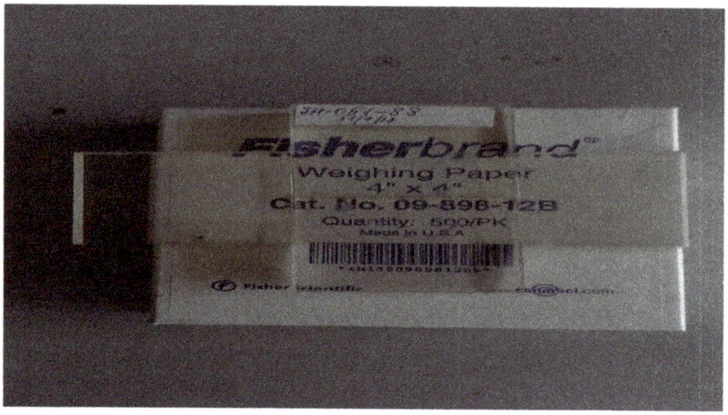

Figure 13. Polished substrates (Ceramtec) edge bonded with glass.

Figure 14. Two polished spinel structures (Ceramtec) edge bonded with glass.

Figures 15 and 16 present the UV-VIS and IR of polished Rafael substrates, some face-to-face glass bonded, and some stacked together with cellotape on the edges, respectively. In Figure 15, the single (-100A), the two substrates stacked together (-100C), and the four substrates stacked together (-100E) have no glass, and their transmittances start at UV at about 200 nm. The single substrate (-100B1) coated with glass (only one face coated) is compared with the single polished substrate (-100A). Comparison shows that coating one face with glass reduced the transmittance in the VIS; the reduction is very small but noticeable. Here, the glass coating was applied to a polished face, and the result was reduction of transmittance in the VIS. When coating was applied to the ground face of semi-polished spinel (Figure 5), the result was an increase in transmittance well above the theoretical value. The two substrates stacked together (-100C) are compared with two substrates face-to-face bonded with glass (-100D1), and (-100D1) has a larger transmittance in the VIS, while (-100C) has a lower transmittance because of four reflections. (-100D1) has a lower transmittance than the single substrate coated with glass (-100B1) in some of the VIS range, but its transmittance in the VIS is high. The four-stacked-together substrates (-100E) are compared with the four substrates that are face-to-face glass bonded (-100F1); transmittance of the glass-bonded substrates (100F1, thickness 3.2 cm, single substrate thickness is 0.8 cm) is much higher than (-100E) in the VIS. Transmittance of (-100F1) is comparable to the transmittance of two substrates glass bonded (-100D1); the glass-bonded substrates behave like a single substrate, i.e., two reflections.

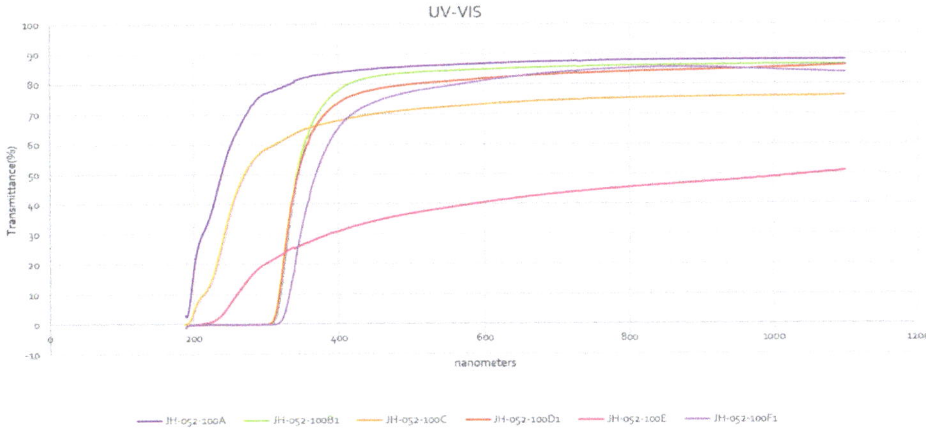

Figure 15. Comparison among transmittances (UV and VIS) of polished substrates, single with no coating, single with coating (one face coated with glass), two stacked together with tape, two bonded with glass, four stacked together with tape, and four substrates face-to-face bonded with glass.

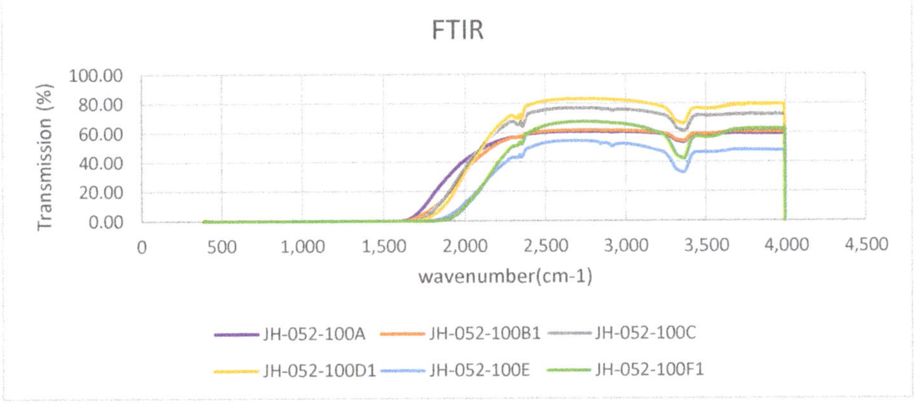

Figure 16. Comparison among transmittances (FTIR) of polished substrates, single with no coating, single with one face coated with glass, two stacked together with tape, two face-to-face bonded with glass, four stacked together with tape, and four substrates face-to-face bonded with three layers of glass.

Figure 16 shows the IR spectra of the substrates of Figure 15. Comparison of the single polished substrate (-100A) with the single coated (only one face coated) substrate (-100B1) shows that they have similar, almost identical, transmittance in the 2400–4000 cm^{-1} range, and below 2350 cm^{-1} the transmittance of the polished uncoated (-100A) substrate is better than that of the one-face-coated polished spinel (-100B1). Comparison of the two polished stacks with cellotape on the edges (-100C) with the two polished, face-to-face glass-bonded substrates (-100D1) shows that -100D1 has better transmittance than -100C. Both show the absorbance at 3380 cm^{-1} (see also Figure 4 for polished substrates; all substrates have this band which is typical to Rafael spinels) and development of new faint band at 3560 cm^{-1}. Comparison of the four polished substrates stacked with cello tape (-100E) with the four face-to-face glass-bonded substrates (-100F1) shows that -100F1 has better transmittance than -100E. As the number of substrates increases the cut-off (where total absorption starts), it shifts to a shorter wavelength.

4. Conclusions

A unique glass composition, which has a matched index of refraction and TCE tailored to be slightly lower than the TCE of spinel to put the glass coating under compression contains ingredients (MgO and Al_2O_3) to slow down the dissolution of spinel ($MgAl_2O_4$) into glass, is non-crystallizing in a broad temperature range and repeated firings, and has good chemical durability, was prepared and described.

Glass coating, when applied to the polished face of spinel, slightly reduced the transmittance relative to polished spinel.

Glass coating, when applied to semi-polished spinel (one face polished the second ground, coating applied to ground face) enhanced the transmittance above the theoretical value of spinel. This new phenomenon first reported here requires further studies to understand how and why the glass coating added a feature of anti-reflecting.

Face-to-face-bonded spinel substrates behave like a single substrate, i.e., have two reflections even for four substrates face-to-face bonded. A thick transparent spinel unit made from face-to-face-bonded substrates was demonstrated.

Funding: This research received no external funding.

Data Availability Statement: Data are contained within the article.

Acknowledgments: The author acknowledges Sofiya Kolusheva for optical measurements, Mariana Dov and Natalia Pears for technical assistance.

Conflicts of Interest: The authors declare no conflict of interest.

References

1. Haggerty, J.S.; Rossetti, M. *Development of Inexpensive Surface Finishing Process for Transparent Armor;* Final Technical Report AD-769 938; Arthur D. Little, Inc.: Cambridge, MA, USA, 1973.
2. Gentilman, R.L.; McGuire, P.T.; Fiore, D.; Ostreicher, K.; Askinazi, J. Low-cost sapphire windows. In *Window and Dome Technologies VIII;* SPIE: Bellingham, WA, USA, 2003; Volume 5078. [CrossRef]
3. Machin, W.S.; Vest, R.W. Reactivity of Alumina Substrates with High Lead Glasses. In *Processing of Crystalline Ceramics;* Materials Science Research; Palmour, H., Davis, R.F., Hare, T.M., Eds.; Springer: Boston, MA, USA, 1978; Volume 11. [CrossRef]
4. Prudenziati, M.; Mortem, B.; Cilloni, F.; Ruffi, G.; Saachi, M. Interactions between alumi-na and high lead glasses for hybrid componenrs. *J. Appl. Phys.* **1989**, *65*, 146–153. [CrossRef]
5. Ganesh, I. A review on magnesium aluminate ($MgAl_2O_4$) spinel, synthesis, processing and applications. *Int. Mater. Rev.* **2013**, *58*, 63–112. [CrossRef]
6. Rubat du Merac, M.; Kleebe, H.J.; Müller, M.M.; Reimanis, I.E. Fifty years of research and development coming to fruition, unraveling the complex interactions during processing of transparent magnesium aluminate ($MgAl_2O_4$) spinel. *J. Am. Ceram. Soc.* **2013**, *96*, 3341–3365. [CrossRef]
7. Johnson, R.; Biswas, P.; Ramavath, P.; Kumar, R.S.; Padmanabham, G. Transparent polycrystalline ceramics: An overview. *Trans. Indian. Ceram. Soc.* **2012**, *71*, 73–85. [CrossRef]
8. Shi, Z.; Zhao, Q.; Guo, B.; Ji, T.; Wang, H. A review on processing Polycrystalline magnesium aluminate spinel ($MgAl_2O_4$): Sintering techniques, material properties and machinability. *Mater. Des.* **2020**, *193*, 108858. [CrossRef]
9. Wang, S.F.; Zhang, J.; Luo, D.W.; Gu, F.; Tang, D.Y.; Dong, Z.L.; Tan, G.E.B.; Que, W.X.; Zhang, T.S.; Li, S.; et al. Transparent ceramics: Processing, materials and applications. *Prog. Solid State Chem.* **2013**, *41*, 20–54. [CrossRef]
10. Krell, A.; Klimke, J.; Hutzler, T. Advanced spinel and submicron Al_2O_3 for transparent armour applications. *J. Eur. Ceram. Soc.* **2009**, *29*, 275–281. [CrossRef]
11. Patterson, M.C.L.; Caiazza, J.E.; Roy, D.W. Transparent spinel development. *Proc. SPIE* **2000**, *4102*, 59–68.
12. Patterson, M.C.L.; Caiazza, J.E.; Roy, D.W. Fabrication of large thick panels of transparent spinel. *Proc. SPIE* **2001**, *4452*, 126–132.
13. Jouini, A.; Yoshikawa, A.; Brenier, A.; Fukuda, T.; Boulon, G.A. Optical properties of transition metal ion-doped $MgAl_2O_4$ spinel for laser application. *Phys. Status Solidi C* **2007**, *4*, 1380–1383. [CrossRef]
14. Goldstein, A.; Loiko, P.; Burshtein, Z.; Skoptsov, N.; Glazunov, I.; Galun, E.; Kuleshov, N.; Yumashev, K. Development of saturable absorbers for laser passive Q-switching near 1.5 mm based on transparent ceramic Co^{2+}:$MgAl_2O_4$. *J. Am. Ceram. Soc.* **2016**, *99*, 1324–1331. [CrossRef]

15. Luo, W.; Ma, P.; Xie, T.; Dai, J.; Pan, Y.; Kou, H.; Li, J. Fabrication and spectroscopic prop-erties of Co:$MgAl_2O_4$ transparent ceramics by the HIP post-treatment. *Opt. Mater.* **2017**, *69*, 152–157. [CrossRef]
16. Liu, Q.; Su, S.; Hu, Z.; Chen, X.; Xie, T.; Yang, Z.; Pan, H.; Liu, X.; Li, J. Fabrication and properties of Co:$MgAl_2O_4$ transparent ceramics for saturable absorber from co-precipitated na-nopowder. *J. Am. Ceram. Soc.* **2019**, *102*, 3097–3102. [CrossRef]
17. Song, E.H.; Zhou, Y.Y.; Wei, Y.; Han, X.X.; Tao, Z.R.; Qiu, R.L.; Xia, Z.G.; Zhang, Q.Y. Thermally stable narrow-band green-emitting phosphor $MgAl_2O_4$:Mn^{2+} toward Wide Color Gamut Backlight Display Application. *J. Mater. Chem. C* **2019**, *7*, 8192–8198. [CrossRef]
18. Rimpongpisarn, T.; Wattanathana, W.; Sukthavorn, K.; Nootsuwan, N.; Hanlumyuang, Y.; Veranitisagul, C.; Laobuthee, A. Novel luminescent PLA/$MgAl_2O_4$:Sm^{3+} composite filaments for 3D printing application. *Mater. Lett.* **2018**, *237*, 270–273. [CrossRef]
19. Chen, C.F.; Doty, F.P.; Houk, R.J.T.; Loutfy, R.O.; Volz, H.M.; Yang, P. Characterizations of a hot-pressed polycrystalline spi-nel:Ce scintillator. *J. Am. Ceram. Soc.* **2010**, *93*, 2399–2402. [CrossRef]
20. Swab, J.J.; Pavlacka, R.; Gilde, G.; Kilczewski, S.; Wright, J.; Harris, D. Determining the Strength of Coarse-Grained AlON and Spinel. *J. Am. Ceram. Soc.* **2013**, *97*, 592–600. [CrossRef]
21. Kilczewski, S.M.; Wright, J.C.; Pavlacka, R.J.; Swab, J.J.; Adams, J.W. *Flexure Strength and Optical Transparency of Magnesium–Aluminate Spinel ($MgAlO_4$): Influence of Polishing and Glass Coating*; Report ARL-TR-7900; US Army Research Laboratory: Adelphi, MD, USA, 2016.
22. Hormadaly, J. Empirical methods for estimating the linear coefficient of expansion of oxide glasses from their composition. *J. Non-Cryst. Solids* **1986**, *79*, 311–324. [CrossRef]
23. Liu, X.; Han, D.; Mao, X.; Zhang, J.; Ren, H.; Lin, H.; Wang, S. Joining transparent spinel ceramics using refractive index–matched glass. *J. Eur. Ceram. Soc.* **2022**, *42*, 3579–3585. [CrossRef]
24. Hormadaly, J. Glass Composition for Coating and Bonding of Polycrystalline Spinel (Transparent Ceramic) Substrates. WO 2023/026276 A1, 2 March 2023.

ceramics

MDPI

Article

Mechanical Behavior of Transparent Spinel Fabricated by Spark Plasma Sintering

Khadidja Hoggas [1,2], Salim Benaissa [3,*], Abdelbaki Cherouana [3], Sofiane Bouheroum [3], Abdenacer Assali [3], Mohamed Hamidouche [1,2] and Gilbert Fantozzi [4]

1 Research Unit on Emerging Materials (RUEM), University of Ferhat Abbas Setif 1, Sétif 19000, Algeria; khadidja.hoggas@yahoo.com (K.H.); mhamidouche@yahoo.fr (M.H.)
2 Institute of Optics and Precision Mechanics, University of Ferhat Abbas Setif 1, Sétif 19000, Algeria
3 Research Unit in Optics and Photonics, CDTA, University of Setif 1, Sétif 19000, Algeria; acherouana@cdta.dz (A.C.); sbouheroum@cdta.dz (S.B.); aassali@cdta.dz (A.A.)
4 Laboratory of Materials Engineering and Sciences (MATEIS), INSA Lyon, 69100 Villeurbanne, France; gilbert.fantozzi@insa-lyon.fr
* Correspondence: sbenaissa@cdta.dz

Abstract: In this work, a transparent nanostructured ceramic magnesium aluminate spinel ($MgAl_2O_4$) was fabricated by Spark Plasma Sintering (SPS) from commercial spinel nano-powders at different temperatures (1300, 1350 and 1400 °C). The sintered samples were thoroughly examined to assess their microstructural, optical, and mechanical properties. Various techniques such as SEM, AFM, spectrophotometer with an integrating sphere, instrumented Vickers indenter, Pin-on-Disk tribometer, scratch tester, and sandblasting device were employed to characterize the sintered samples. The results indicated the significant impact of the sintering temperature on the properties of the spinel samples. Particularly, the samples sintered at T = 1350 °C exhibited the highest Real In-line Transmission (RIT = 72% at 550 nm and 80% at 1000 nm). These samples demonstrated the highest hardness value (HV = 16.7 GPa) compared to those sintered at 1300 °C (HV = 15.6 GPa) and 1400 °C (HV = 15.1 GPa). The measured fracture toughness of the sintered samples increased substantially with increasing sintering temperature. Similarly, the tribological study revealed that the friction coefficient of the sintered spinel samples increased with the sintering temperature, and the spinel sintered at 1350 °C exhibited the lowest wear rate. Additionally, sandblasting and scratch tests confirmed the significant influence of the sintering temperature on the mechanical properties of the fabricated spinels. Overall, the spinel sintered at 1350 °C presented the best compromise in terms of all the evaluated properties.

Keywords: spinel; SPS; transparent ceramics; optical properties; mechanical properties; sandblasting

1. Introduction

The most commonly used transparent ceramics in industries today are single crystals and glasses. However, although highly transparent, they have rather poor mechanical properties (mechanical strength, hardness, wear resistance). In addition to being expensive, single crystals with complex shapes are hard to obtain. It is therefore important to find other appropriate alternatives. In this context, transparent polycrystalline nanostructured ceramics can be candidates, as they can offer very interesting combinations, such as: Opto-Mechanical properties, easier shaping and a possibility of large-scale production [1,2]. They are also characterized by a good thermal resistivity, a high chemical inertness and a high doping rate of active ions. These characteristics allow controlling of their optical performance. They are also optically isotropic and present a wide range of UV–Visible–IR transmission. Thanks to these favorable physical, chemical and mechanical properties, transparent ceramics can be exploited in a wide range of applications, including laser materials, scintillators, optical lenses, and even for creating transparent armor. [2]. The manufacturing of polycrystalline ceramics with maximum optical transparency requires the elimination of all causes of light losses, such as light scattering from pores and grain boundaries. Therefore, it is necessary to use highly pure raw materials with the finest possible grain size to ensure high densities (>99.9%) and small pore sizes [3]. To meet these criteria, two main families of sintering can be implemented: (a) sintering with the application of an external macroscopic stress, including hot pressing (HP), hot isostatic pressing (HIP) and Spark Plasma Sintering (SPS), or (b) sintering without external macroscopic stress (microwave sintering).

In the last years, SPS has been shown to be a reliable alternative technique for manufacturing transparent polycrystalline ceramics at relatively low temperatures in shorter times [4]. The principle of SPS sintering is based on the application of an associated uniaxial pressure with a pulsed direct electric current. In fact, several studies confirm that the control of SPS parameters, especially heating rate, sintering temperature, holding time and pressure, allows the fabrication of fine-grained dense transparent ceramics with good optical and mechanical properties [5–9]. The application of high pressure, for example, during SPS sintering, allows good control of grain growth at lower temperatures and for shorter holding times. Sokol et al. [10] showed that the application of high pressure results in a fully dense transparent spinel, which exhibits a unique combination of high real in-line transmission (>80% at 500 nm), high Vickers hardness (up to 20 GPa) and fine grain size (~30 nm). The SPS technique has been used for many ceramic materials, namely, ZrO_2 [11], Al_2O_3 [12], $Y_3Al_5O_{12}$ [13], and MgO [14] and $MgAl_2O_4$ [5]. $MgAl_2O_4$ spinel is one of the promising transparent ceramics; it is difficult to fabricate directly from high-purity precursor powders using conventional pressureless sintering techniques [15]. However, it can be easily and qualitatively fabricated by hot pressing techniques, mainly the SPS [16]. It is used for several applications, such as transparent windows, domes, armor and optical refractories [7]. This is thanks to its important properties such as high transparency over a wide range of wavelengths (0.2–5.5 μm), relatively low density (3.58 g/cm^{-3}), a high level of hardness (between 16–20 GPa), high mechanical strength (150–315 MPa), high melting temperature (2135 °C) and high electrical resistivity [7].

In a potentially aggressive environment such as the Sahara, erosion of glass caused by windborne sand is a serious problem that can affect its transparency. In such conditions, spinel may be more suitable than glass for armor and protective window applications, as these types of applications require high optical and mechanical performance. The study of the resistance of a material to sand erosion can be effectively simulated in the laboratory using a sandblasting system.

Studies have shown that material removal during the erosion of brittle materials occurs primarily through the formation and propagation of radial and lateral cracks, which cause the ejection of grains from the material surface [17]. This material removal depends on several factors, such as the properties of the eroding material (shape, size, hardness, etc.) and the eroded material (hardness, toughness, etc.) as well as the conditions of exposure to erosion (speed, mass, angle of attack, etc.) [18–20]. Unfortunately, there have been only a few studies in the literature dealing with the erosion behavior of transparent ceramics. Lallemant et al. [21] and Von Helden [22] have studied the sand erosion of transparent ceramics such as alumina, $MgAl_2O_4$ spinel and different types of glass. Their results revealed that there is a strong relationship between the microstructure of the material and its resistance to erosion and alteration of its optical transmittance. Similarly, Zhou et al. [23] have found that the erosion resistance of alumina increased with the increase in hardness and fracture toughness. On the other hand, Curkovic et al. [24] showed that the erosive wear behavior of transparent alumina ceramic depends on the eroding properties (shape and hardness). Furthermore, they found that the maximum erosion occurs at an impact angle of 90°.

From all the forgoing, a transparent optical ceramic must have high mechanical properties, including sand erosion resistance and scratch resistance, in order to maintain good surface quality and high transparency. The aim of this work is the fabrication of a dense $MgAl_2O_4$ spinel with high optical transmission and high mechanical properties for Sahara applications. For this purpose, SPS sintering of commercial spinel nano-powders at different temperatures was performed and the effects of sintering temperature on the microstructure, optical and mechanical properties of the fabricated $MgAl_2O_4$ spinel were studied. The scratch resistance and the ability to maintain good surface quality and transmission after sandblasting of the fabricated spinel are examined.

2. Materials and Methods

2.1. Sintering of Samples

A nano-powder of magnesium aluminate spinel $MgAl_2O_4$ (S25CRX 14) commercialized by Baikowski Company (La Balme-de-Sillingy, Poisy, France) was used as a raw material. Table 1 shows the impurity amounts according to the supplier's data sheet. This small impurity content has practically no effect on the properties of a transparent spinel.

Table 1. Impurities in the $MgAl_2O_4$ spinel powder.

Component	Na	K	Fe	Si	Ca	S
Amount (ppm)	11	13	6.5	14	6.9	300

The samples were sintered using an SPS sintering machine, type FCT System HP D25, Rauenstein, Germany. A quantity of 3.5 g of spinel powder was introduced into a graphite die with an internal diameter of 20 mm. Pellets were produced under a pressure of about 73 MPa with sintering temperatures of 1300, 1350 and 1400 °C. The heating rate was 100 °C/min up to 800 °C, then 10 °C/min up to 1100 °C and 1 °C/min up to the final sintering temperature. The change in heating rate during the SPS sintering process can have a significant influence on the density and development of the microstructure of the consolidated material. A high heating rate can lead to insufficient densification which is detrimental to the optical and mechanical properties, while too low a heating rate can extend the sintering process excessively. Thus, finding the right equilibrium of heating rate is essential to achieve the desired properties of the final material as shown by Morita et al. [25]. This sintering cycle of spinel powder was chosen based on the study of Bonnefont et al. [26] and Benaissa et al. [27].

Before characterizing the sintered pellets, both sides of the sample were polished in order to eliminate geometric imperfections and the layer contaminated by carbon of the graphite die and punch. The polishing procedure used in this study was identical to that reported in the study of [3].

2.2. Analysis and Characterization Techniques

The microstructures of the samples were determined using a scanning electron microscope (SEM, JSM-7001F, JEOL, Akishima, Tokyo, Japan.). The grain size was also determined from SEM micrographs using ImageJ software with a correction factor equal to 1.22 [28]. The surface topography examination was carried out using an atomic force microscope (AFM) from Asylum Research, an Oxford Instruments company, type MFP-3D, Santa Barbara, CA, USA. A Leica DCM 8 laser scanning 3D confocal microscope (the state of Hesse in central Germany) was used to measure the surface roughness (root mean square of the roughness profile (RMS)). The sample density was determined by the Archimedes method. The optical transmission was measured in the range of 200 to 1200 nm using a Shimadzu UV-1800 spectrophotometer (Kyoto, Japan). The RIT for the sample with thickness (d_2 = 0.88 mm), sintered at three different temperatures, was estimated using the Apetz model [28]:

$$RIT(d_2) = (1 - Rs)\left(\frac{RIT(d_1)}{1 - Rs}\right)^{\frac{d_2}{d_1}} \quad (1)$$

where R_S is the total normal surface reflectance (~0.14) and RIT (d_1) is the real in-line transmission for the real thickness of the sample.

The total transmission (TFT), the diffuse transmission (TD), the specular reflection (*Rs*), the total reflection (RT), and the diffuse reflection (RD) were measured by a spectrophotometer type Jasco-670 (Hachioji, Tokyo, Japan), equipped with an integration sphere. The hardness measurements were performed by the instrumented Vickers indentation method, using a Zwick Roell-type apparatus, ZHU2.5 (Ulm, Germany). The Vickers hardness values obtained represent the average of three measurements taken on the samples. Vickers hardness (GPa) was calculated by the following formula [29]:

$$H_V = 0.0018544 \times (F/(d)^2) \quad (2)$$

where F is the applied indentation load [N] and d is the average length of the diagonals of the indentation imprint [mm].

Young's modulus was determined using ultrasonic testing in a single experiment [30]. The fracture toughness (K_{IC}) of the transparent spinel was measured using the indentation technique on specimens with polished surfaces, with six measurements taken on samples under a 5 N load. The measured radial crack lengths (2C in m) were used to determine the fracture toughness by the equation of Anstis et al. [31], which recently concluded that that the following formula provides the most accurate characterization of ceramics. [32]:

$$K_{IC} = 0.016(\frac{E}{H})^{1/2} \times \left(\frac{F}{C^{3/2}}\right) \quad (3)$$

where H is the Vickers hardness (GPa) and E (GPa) is the elastic modulus.

The tribological tests were carried out using a CSM-type Pin-On-Disk Tribometer (Version 4.5.Q, Peseux, Switzerland) with an alumina ball (Ø 6 mm) as a pin rubbing against sintered samples as disks.

Each sample was rubbed under a normal force of 10 N with a sliding radius of 6.2 mm, a linear speed of 100 rpm and a test duration of 15 min. All the tribological tests were performed in a dry environment at a room temperature of 20 °C and humidity of ~40%. During all the tests, the friction coefficient was measured as a function of time. In order to evaluate the wear rate, the width of the wear track was measured using an optical transmission microscope.

For the sandblasting tests, the transparent spinels were eroded using a horizontal type of sandblasting device [33]. The latter complies with DIN 50 332 standards and ASTM G76. A natural sand from the region of El-Oued (Southeast of Algeria) was used as eroding material. Figure 1 showed that this sand has various grains with different shapes and with an elongated index (Ei = Dmax/dmax) of 1.44. It has a fairly homogeneous particle size ranging between 250 and 500 μm with an average particle size of 374 μm (Figure 2). The chemical composition of this sand was indicated in Table 2. The fabricated spinel samples were sandblasted on one side under the following conditions:

- Projected mass: m = 200 g
- Air flow speed: v = 30 m/s
- Angle of impact: α = 90°
- Distance between the nozzle outlet and the sample: x = 50 mm.

Table 2. Chemical composition of the eroding sand used.

Oxides	SiO_2	Al_2O_3	CaO	K_2O	SO_3	Fe_2O_3	MgO	SrO
Amount (wt.%)	90.8	3.86	1.36	1.32	1.22	0.592	0.459	0.0140

Figure 1. Micrograph of the sand particles used as projectiles impacts.

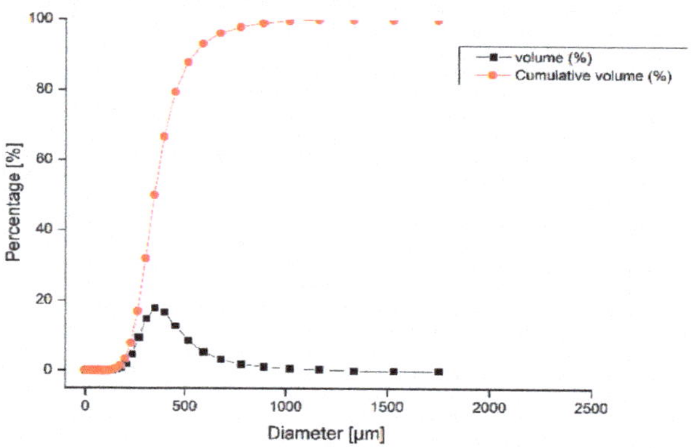

Medium diameter	374.66 µm
D (0.1)	236.72 µm
D (0.5)	344.63 µm
D (0.9)	548.55 µm

Figure 2. Particle-size distribution of the eroding sand used.

The difference in surface roughness of the sample before and after sandblasting may reflect its rate of damage following its exposure to erosion. For this purpose, roughness parameters were measured before and after sandblasting, using a Leica DCM 8 3D laser scanning confocal microscope. Two main parameters were chosen to evaluate the roughness based on this test: the root mean square roughness (Rq) according to ISO 4287 and the root mean square height (Sq) according to ISO 25178.

The progressive linear scratch test was performed using a CSM Revetest scratch device equipped with a Rockwell diamond indenter with a radius of 200 µm under a progressively increasing load from 1 to 60 N. The loading rate was 59 N/min. The scratch length was 5 mm. During all the tests, the penetration depth, the applied load and the acoustic emission were recorded to assess the characteristics of cracks and determine the critical forces. Finally, the scratches were examined by light microscopy.

3. Results and Discussion

3.1. Microstructural Characterization

SEM micrographs of the spinel samples sintered at 1300, 1350 and 1400 °C are presented in Figure 3a–c. It is clear that all the samples have a dense and homogeneous microstructure. However, it is found that the sintering temperature has a significant effect on the microstructure of the fabricated spinel, where the grain size of the samples increased significantly with increasing sintering temperature. The sample sintered at 1300 °C has the smallest average grain size of around 685 nm; this size increased to 1.9 µm in the sample sintered at 1350 °C, and reaches its maximum in the sample sintered at 1400 °C, which has the largest average grain size of about 5.3 µm. The defects that appeared on the images of

1350 and 1400 °C perhaps were caused only by the chemical treatment that we carried out before the observation at the SEM, as shown by Benaissa et al. [26].

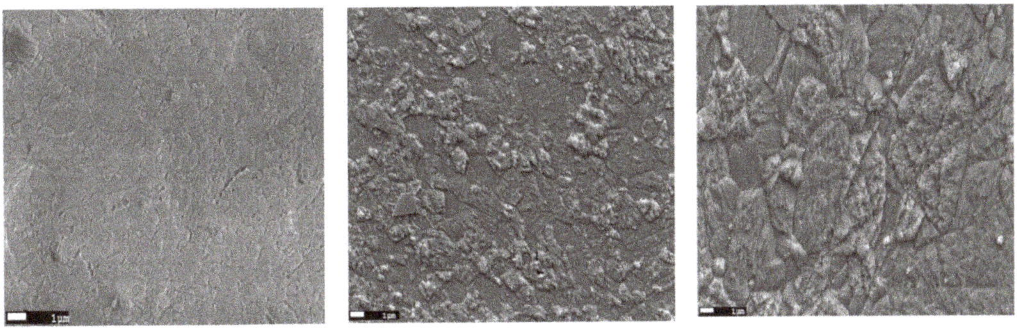

(a) (b) (c)

Figure 3. SEM micrographs of the samples sintered by SPS at: (**a**) 1300 °C; (**b**) 1350 °C and (**c**) 1400 °C.

The 2D and 3D surface morphologies of the samples sintered at different temperatures are presented in Figure 4. The effect of the sintering temperature is clearly visible on the morphology of the fabricated samples, where the grain size increased with the increase in the sintering temperature, which confirms the SEM observations (Figure 3). Furthermore, the average surface roughness RMS values obtained by AFM, in the same scanning area (25 μm^2), were 5.8 ± 0.74, 8.2 ± 2.03 and 61.9 ± 5.06 nm for the samples sintered at 1300, 1350 and 1400 °C, respectively. The increase in roughness can be mainly explained by the increase in grain size by coalescence of neighboring grains due to the increase in sintering temperature.

The relative density of the sintered samples was determined according to the Archimedes principle. The density values were 99.65 ± 0.1%, 99.74 ± 0.11% and 99.86 ± 0.03% for the samples sintered at 1300, 1350 and 1400 °C, respectively. The obtained values are the means of five measurements per sample. The slight increase in density with increasing sintering temperature could be explained by the coarsening of grains and promoting of densification by sliding and rearrangement of grains, because the grains bind or weld together as they grow, which can lead to consolidation of the material. Consolidation can cause the relative density to increase because the grains are more tightly bound and the space between them is reduced [34].

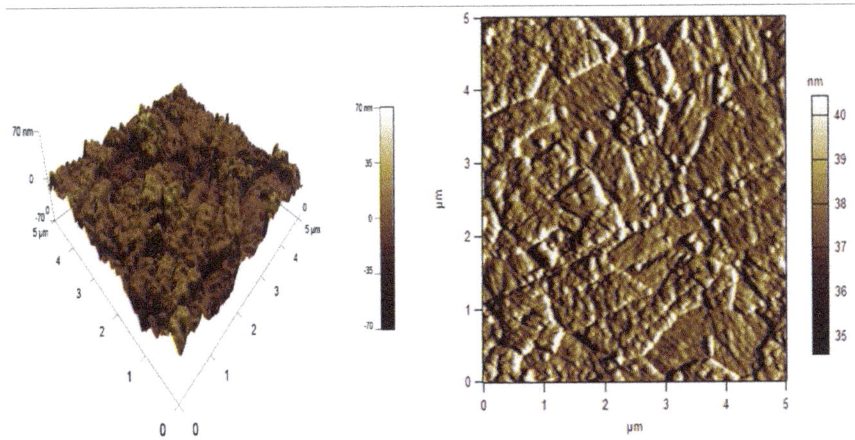

(a)

Figure 4. *Cont.*

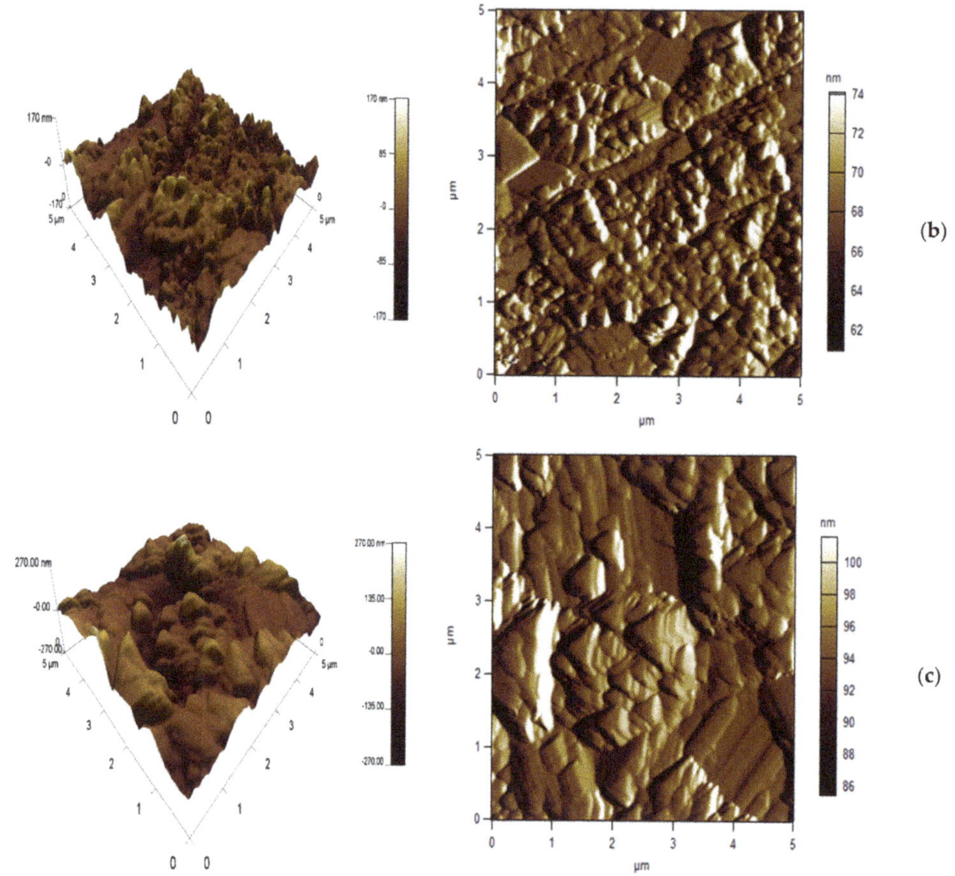

Figure 4. 2D and 3D AFM images of the samples sintered at: (**a**) 1300 °C, (**b**) 1350 °C and (**c**) 1400 °C.

3.2. Optical Characterization

Figure 5 shows the spectra of real in-line transmission of the samples sintered at different temperatures. It can be observed that the samples sintered at 1300 and 1350 °C have almost the same high RIT in the visible range (400–800 nm) and that the best optical real in-line transmission is reached with samples sintered at 1300 and 1350 °C. This RIT decreased abruptly in the sample sintered at 1400 °C, which has a very low transmission. The decrease in the transmission at high temperature can be explained by the strong contamination by carbon during sintering; Hammoud et al. [35] showed that a contamination occurred by carbon clusters originating from the powder and the environment of the SPS (graphite foil, die and felt). A high SPS temperature may further favor the diffusion from the carbon clusters, which explains the increased light absorption observed in samples sintered at 1400 °C. However, the carbon contamination can be reduced even slightly by a thermal annealing under air, as shown by Zegadi et al. [3]. Not only carbon contamination but also grain size may be an important parameter influencing transmission, as shown by Rothman et al. [36]. It should be noted that our results regarding transmission as a function of SPS temperature are in good agreement with those in the literature [37,38].

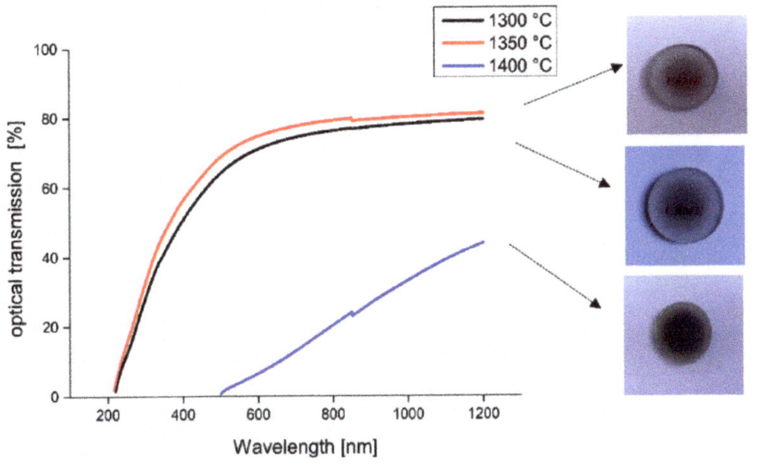

Figure 5. The Real In-line Transmission spectra of samples sintered by SPS at 1300, 1350 and 1400 °C.

Table 3 summarizes the RIT values recorded at 550 and 1000 nm wavelengths after correction of sample thicknesses to 0.88 mm. It can be noticed that the samples sintered at 1350 °C have the highest value of transmission (RIT = 72% at 550 nm and RIT = 80% at 1000 nm). These results are almost higher than those obtained by Benaissa et al. [27], who found an RIT = 45% at a wavelength of 550 nm and RIT = 67% at 1000 nm by SPS sintering of S30CR at the same temperature of sintering.

Table 3. RIT of the $MgAl_2O_4$ spinel sintered by SPS at different temperatures (d_2 = 0.88 mm).

SPS Temperature	1300 °C	1350 °C	1400 °C
RIT (%) at 550 nm	68.1	72.3	3.4
RIT (%) at 1000 nm	78.2	80.2	33.2

To better understand the effect of sintering temperature on the different optical properties of the fabricated transparent spinel, we present the intensity of transmitted, reflected, absorbed and scattered light in Figure 6 as a function of three different wavelengths. The first conclusion that can be drawn is that the RIT increased with the increase in the wavelength, while the absorbed intensity, the diffuse transmission and the diffuse reflection decreased. The second conclusion is that the sample sintered at 1350 °C exhibited the best transmission at any wavelength, whereas the sample sintered at 1400 °C exhibited the lowest transmission at any wavelength. The poor optical transmission of the sample sintered at 1400 °C can be explained by carbon contamination during sintering, where this sample has been identified as a potential cause of a decrease in optical properties [35,39,40].

3.3. Mechanical Characterization

The variation of Vickers hardness of the sintered specimens as a function of indentation load is illustrated in Figure 7. It can be clearly seen that the hardness of the samples sintered at 1350 °C is higher than of those sintered at 1300 and 1400 °C, regardless of the indentation load. The observed variation of the hardness of the samples sintered at different temperatures is attributed mainly to the variation of their porosity and grain size (Hall and Petch law) [10]. This is because the reduction in grain size limits the propagation of dislocations, which strengthens the material and leads to an increase in its hardness. Conversely, samples with larger grain sizes demonstrate the lowest hardness values. The larger grain size promotes dislocation propagation, weakening the material and leading to a decrease in hardness.

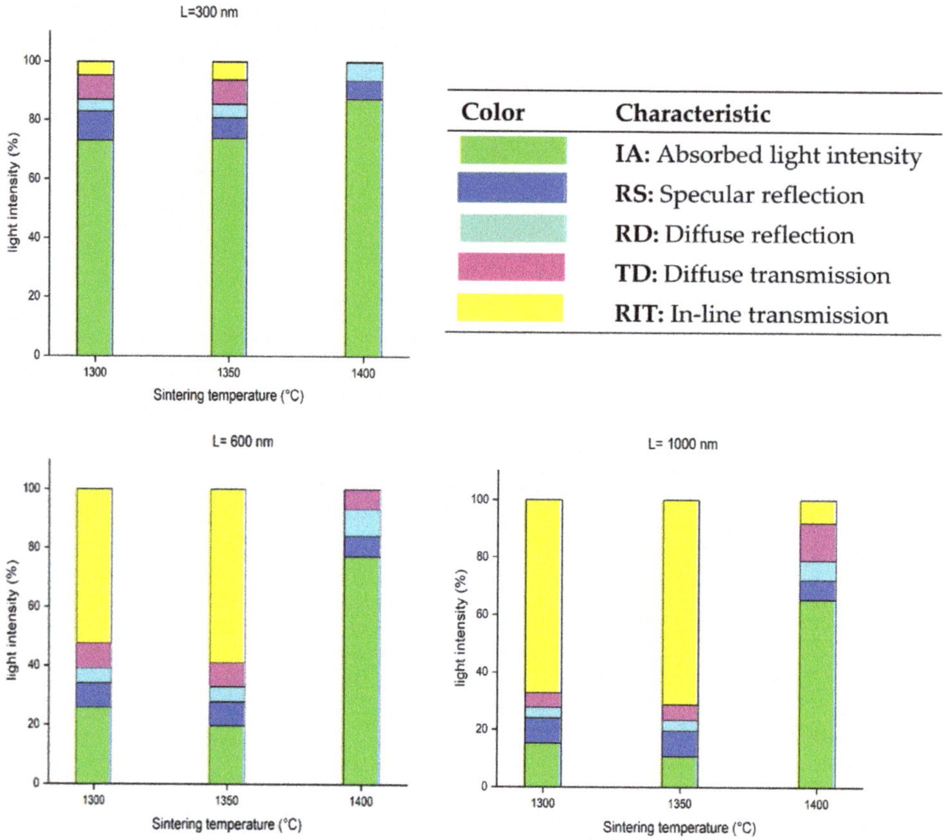

Figure 6. Distribution of incident light intensity within the fabricated spinel pellets.

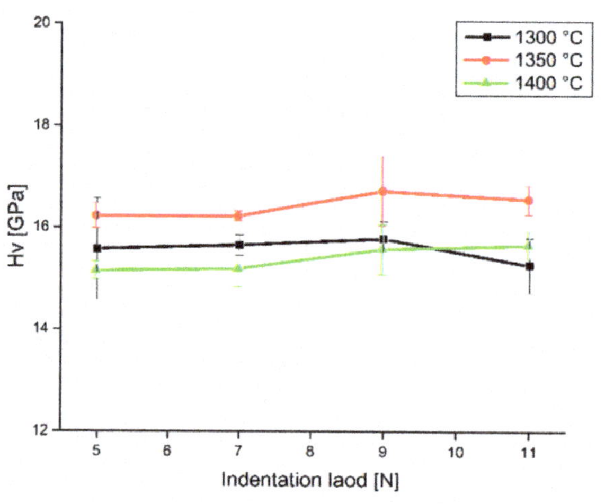

Figure 7. Vickers hardness as function of indentation load of the spinel sintered at 1300, 1350 and 1400 °C.

It should be noted that, in the case of spinel, the homogeneity of the microstructure, in particular the grain size, could have a more important effect on the mechanical properties than the porosity rate and the presence of inclusions [41]. From Figure 7, it can also be seen that the hardness values do not depend on the indentation load; this is due to the uniformity of the microstructure and the high relative density of the prepared spinel pellets. On the other hand, the obtained Vickers hardness values were always higher than 15.5 GPa, regardless of the sintering temperature. It should be noted that this hardness value with the same measurement method is slightly higher than that found by Krell et al. [42] (HV = 15 GPa), and considerably higher than that found by Benaissa et al. (H_V = 15 GPa at 1350 °C) [27].

Table 4 show the results of elastic values obtained by the ultrasonic method for the spinel samples sintered at 1300, 1350 and 1400 °C. A gradual increase in Young's modulus with the increase in sintering temperature can be clearly observed, where the elastic modulus goes from 256 to 285 GPa by increasing the sintering temperature from 1300 to 1400 °C. This increase can be explained mainly by the microstructural changes and the increase in density with the increase in the sintering temperature, where it is well known that the elastic modulus is sensitive to the porosity.

Table 4. Elastic modulus and fracture toughness of the spinel samples.

Sintering Temperature	1300 °C	1350 °C	1400 °C
Elastic modulus [GPa]	265	272	285
Fracture Toughness [MPa$\sqrt{m}$]	1.5 ± 0.1	1.7 ± 0.3	2.5 ± 0.4

The results of this study indicate that in order to improve the mechanical properties of spinel, a homogeneous and fine microstructure must be maintained. However, in the case of low temperature sintering (1300 °C), spinel has a slight porosity and a fine microstructure. In contrast, spinel sintered at intermediate temperature (1350 °C) is dense with a homogeneous microstructure and slightly increased grain size. Finally, in the case of sintering at a high temperature (1400 °C), the spinel presents a coarse microstructure.

The fracture toughness (K_{IC}) calculated from Equation (3) for an indentation load of 5 N was presented as a function of sintering temperature in Table 4. It can be noticed that the values of the fabricated spinel pellets increased gradually from 1.5 MPa$\sqrt{m}$ for 1300 °C to 2.5 MPa$\sqrt{m}$ when the samples were sintered at the temperature of 1400 °C.

This increase in fracture toughness is mainly due to the increase in grain size with increasing sintering temperature. In fact, the improved grain bridging increases with the grain size; this grain bridging is behind the improvement of the fracture toughness of the fabricated ceramic. It should be noted that our best fracture toughness value of the sample sintered at T = 1400 °C (K_{IC} = 2.5 MPa$\sqrt{m}$) was higher than that reported by Tokariev et al. (K_{IC} = 1.8 MPa$\sqrt{m}$) [43] and Benaissa et al. (K_{IC} = 2.2 MPa$\sqrt{m}$) [27]. Moreover, it was slightly lower than that reported by Von Helden (K_{IC} = 2.8 MPa$\sqrt{m}$) [22] under almost the same test conditions. This proves that our prepared spinel specimens exhibit a very accepted fracture toughness.

Figure 8 shows the evolution of the friction coefficient as a function of time for the spinel samples sintered by SPS at different temperatures. It can be observed that the sintering temperature has a significant influence on the tribological behavior of the fabricated spinel. The sample sintered at 1350 °C presented the most stable and the lowest coefficient of friction with a value of about 0.35. The friction coefficient increased to about 0.42 and became less stable in the sample sintered at 1300 °C.

The sample sintered at T = 1400 °C presented the most fluctuating and the highest friction coefficient with an average value of about 0.47. Thus, it can be concluded that the spinel sample sintered at 1350 °C exhibits the best tribological behavior of the friction reduction among the fabricated samples. The friction coefficient is an extrinsic material property that can be influenced by the conditions surrounding the material (temperature, humidity, lubrication, etc.) as well as the intrinsic material properties such as hardness, roughness and especially the phase composition of the material. However, hardness and roughness seem to be the most influential factors on the friction coefficient in this study, where the sample sintered at 1350 °C, which has the highest hardness, and relatively low roughness, has the lowest friction coefficient value and showed the best frictional behavior.

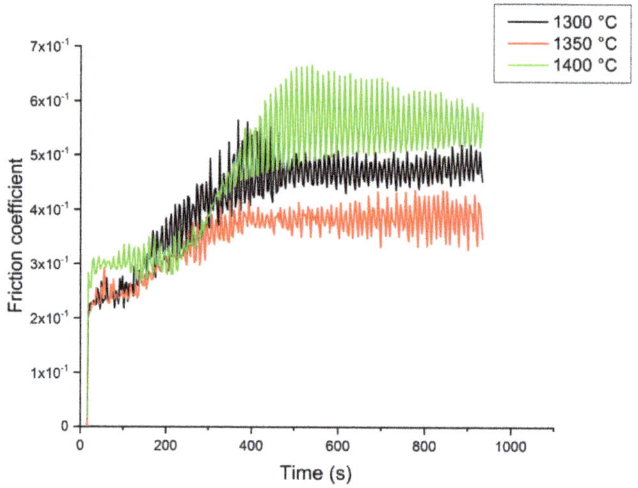

Figure 8. Evolution of friction coefficient as a function of time for the spinel sintered at different temperatures.

These results show that the friction coefficient depends on the one hand on hardness, and on the other hand on roughness linked to the grain size. The sample sintered at 1400 °C exhibits a large grain size (5.3 µm) and therefore a high roughness (62 nm) and the lowest hardness, giving a high friction coefficient. These results confirm those found by other researchers [27].

The width of the wear tracks can be an indication about the wear rate and the wear resistance of the materials subjected to rubbing. Figure 9 illustrates the variation in the width of the wear tracks of the samples sintered by SPS at different temperatures that underwent the wear test. The values presented are the average results of seven measurements, taken with the utmost care to ensure consistency in the experimental conditions and to ensure the reliability of the data. It is evident that the sample sintered at 1350 °C exhibits the narrowest wear track width (116 µm), albeit with a relatively high standard deviation, which could be attributed to potential surface flatness defects or inadequate cleaning of the ball during the test. The sample sintered at 1400 °C, in contrast, exhibited the widest wear (133 µm) track and the lowest wear resistance. This tribological behavior of our manufactured spinels can be explained firstly by the variation in the friction coefficient, where the spinel sintered at 1350 °C has the lowest friction coefficient and showed the highest wear resistance, while that sintered at 1400 °C has the highest friction coefficient and showed the lowest wear resistance. The tribological behavior of our materials can also be explained by the hardness variation, where the material with the highest hardness value presented the lowest wear rate, and vice versa, this corresponds to the Archard wear equation [44], which demonstrates an inverse relationship between the hardness of a material and its rate of wear.

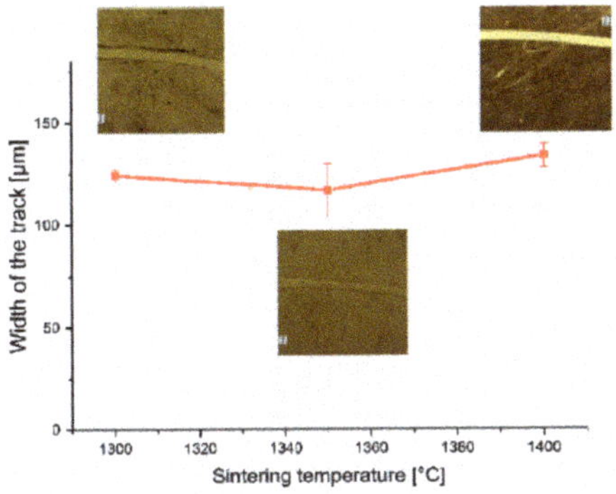

Figure 9. Variation in the track width as a function of sintering temperature.

3.4. Sandblasting

The erosion behavior of brittle materials such as polycrystalline ceramics is controlled by the formation of lateral cracks parallel to the surface. These cracks easily propagate to form a network of cracks propagating through grain boundaries. Repeated shocks by sand particles will easily remove surface material with grain ejection from the top layer. The volume of the removed material depends on the material fracture toughness and hardness [23]. As these latter values are different for each sample (Figure 7 and Table 4), one would expect different erosion rates. For the microscopic observations of defects, an optical microscope is used. It allows the surface defects to be observed and photographed and then the damage rate of a transparent ceramic surface to be seen (Figure 10). From these micrographs, it can be noticed that the sample's surface is completely eroded, and the sandblasting defects resulting from micro-fragmentation, particularly lateral cracks, exhibit a random distribution across the entire exposed surface. In addition, it is noticed that the damage worsens as the sintering temperature decreases. Thus, it is found that the spinel sintered at 1300 °C is more damaged compared to the sample sintered at 1400 °C, due to its low fracture toughness. The damage rate corresponding to the samples sintered at 1300, 1350 and 1400 °C is equal to 46.7%, 34.6% and 32%, respectively. In order to examine the mechanism of erosion and surface damage, the top view of uneroded and eroded surfaces of sintered samples was observed by confocal microscope (Figure 11). This finding confirms that, after sandblasting, the ejection of the grains from the surface layer leaves hollows in the samples surface and increases the roughness, which subsequently influences the surface reflectance Rs, as already mentioned in literature [24]. It is also noticed that the samples sintered at 1400 °C have deeper defects than the other samples because of the coarse microstructure (grain size of around 5.3 µm), and the surface state becomes more and more damaged and rough.

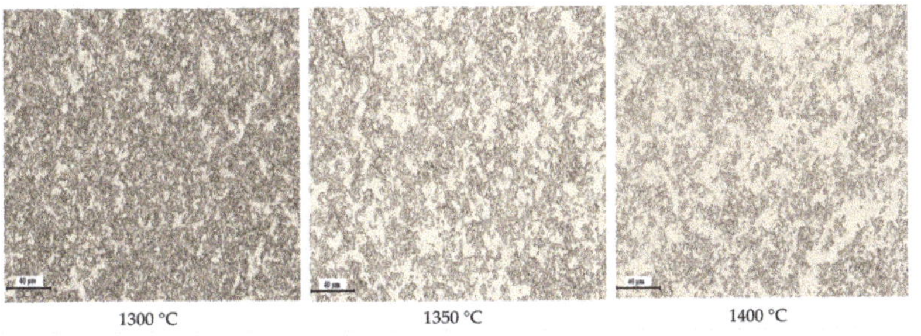

Figure 10. Microscopic observations of the surface defects for three sintering temperatures after sandblasting.

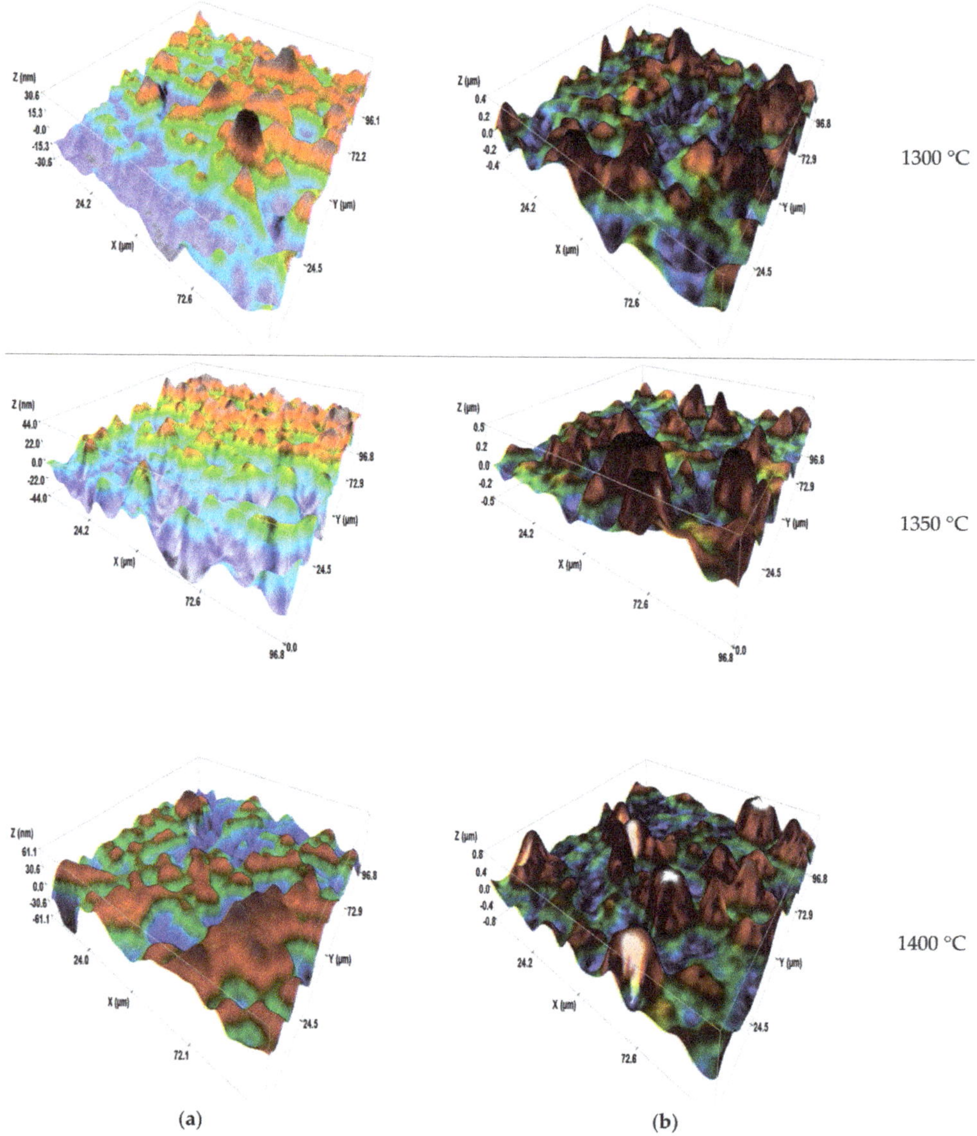

Figure 11. 3D images of confocal microscope for sintered sample surfaces, (**a**) before and (**b**) after the erosion.

The changes in the roughness value quadratic height (Sq: extended surface roughness parameter relating to the area) of different transparent samples before and after erosion are shown in Figure 12. It is observed that the roughness in the raw state is low in the case of samples sintered at 1300 and 1350 °C (75 and 73 nm). These results show that the samples undergo good polishing with fine microstructures; except in the case of spinel sintered at 1400 °C, which has a high Sq of the order of 125 nm. Furthermore, when the roughness results after sandblasting are examined, it can be noticed that the values of Sq increase from 75 nm up to 134 nm for the sample sintered at 1300 °C. For spinel sintered at 1350 °C, the Sq reaches 139 nm. As for spinel sintered at 1400 °C, the roughness reaches 167 nm.

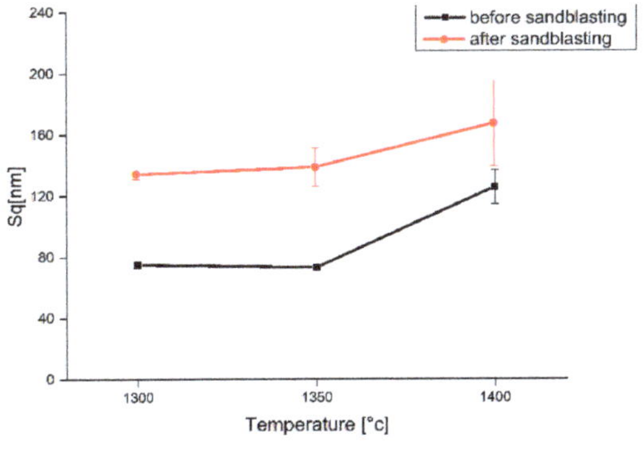

Figure 12. Variation of the roughness of the sintered samples before and after sandblasting.

The surface state is influenced by the defects generated by the grains of sand and the microstructure. In this case, it is found that the spinel with a larger microstructure (sintered at 1400 °C) exhibits higher roughness values than the sintered spinel with a fine microstructure (sintered at 1300 °C). The microstructure effect on the morphology of eroded surface can be explained by treating the sintered sample's surface with sand particles. The removed grains leave marks corresponding to their sizes. As a result, the roughness of the sanded surface increases with the increase in size of the removals. In a study subsequently carried out by Mroz and his collaborators [45] on the erosion of the nanostructured spinel ($MgAl_2O_4$) by sand and rain, the nano-spinel with a grain size of around 345 nm exhibits better resistance to erosion when compared with the fine grain-size spinel of 970 nm.

Figure 13 shows the optical transmission spectra of transparent spinel eroded by a mass of sand of 200 g. The spinel spectra have the same appearance, but the level of the curves differs. This shift is due to the degree of the surface damage, which scatters the incident light. It is clearly seen that the initial state (before sandblasting) is placed at the highest level regardless of the sintering temperature. After sandblasting the samples show a drop in optical transmission of about 7% at λ = 550 nm for the sintering at 1300 °C and 1350 °C. This loss of optical quality is due to the increased reflection of light from the surface, which is caused by defects induced by sand impacts.

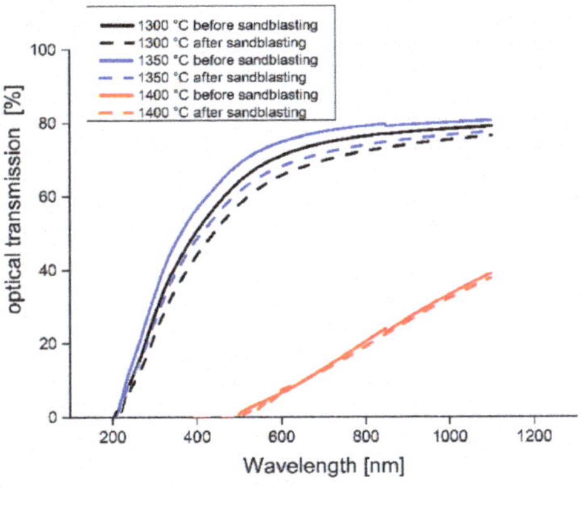

Figure 13. Variation of optical transmission of sintered spinel at different temperatures before and after sandblasting.

During the progressive scratching tests, it was noticed that when there is an increase in the applied normal load, three successive damage regimes occur [46] (Figure 14): first, a micro-ductile regime where there is a permanent furrow without visible damage (a); second, a cracking regime which is characterized by the initiation of a first radial crack under the effect of a normal critical load (b); and third, a micro-abrasive regime, caused by the normal breaking load (c). In this latter case, the damage becomes more important, and the formation of strong lateral cracks induces spalling, which can cause a removal of material.

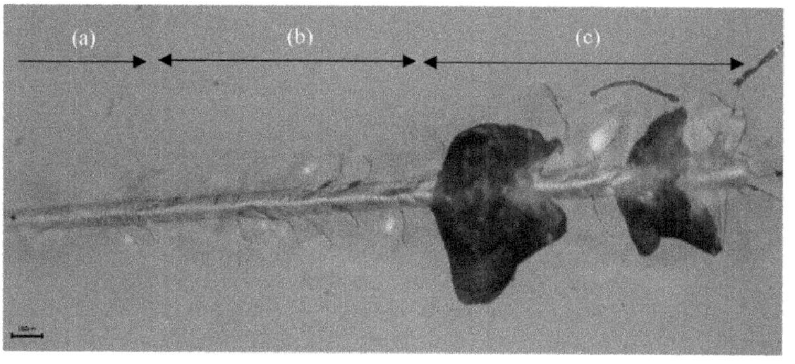

Figure 14. Microscopic observations of the scratching of spinel sintered at 1300 °C: (**a**) Micro-ductile regime; (**b**) Cracking regime; (**c**) Micro-abrasion regime.

From the optical observations, the normal critical cracking loads and the normal critical damage loads for the samples sintered at different temperatures are determined (Table 5).

Table 5. Critical forces of the scratch test for the different sintered samples.

Critical Force	FC1 (N)	FC2 (N)
1300 °C	35.2	49.8
1350 °C	30.9	42.5
1400 °C	19.5	34.5

Sample S25CRX 14 sintered at 1400 °C has lower critical load values than the other two samples sintered at 1300 and 1350 °C. This behaviour is explained by the lower hardness of the sample sintered at a higher temperature due to the less fine-grained microstructure. Figure 15 illustrates the scratching behaviour of the different samples according to the sintering temperature. The increase in scratching force leads to large spalling areas when the sintering temperature is higher (1400 °C). The primary damage mechanism during spinel scratching is a brittle fracture. This type of behaviour has also been observed in the case of alumina [47–49].

During the scratch test, the system records the depth of penetration (Pd) and the acoustic signals (AE) associated with cracking or chipping. Figure 16a–c presents the variation of penetration depth and the acoustic emission recorded during scratching of transparent spinels, sintered at different temperatures. When the microstructure is finer, and sintering is at 1300 °C, the beginning of the acoustic emission takes place at a higher normal load. As the applied load increases and spalls form, the emission signals become more intense. Indeed, as the normal load increases, the residual penetration depth increases. At the end of the scratch test, we noted the depth reached as follows: 20.6, 7.5 and 14.2 μm, respectively, for the samples sintered at 1300, 1350 and 1400 °C.

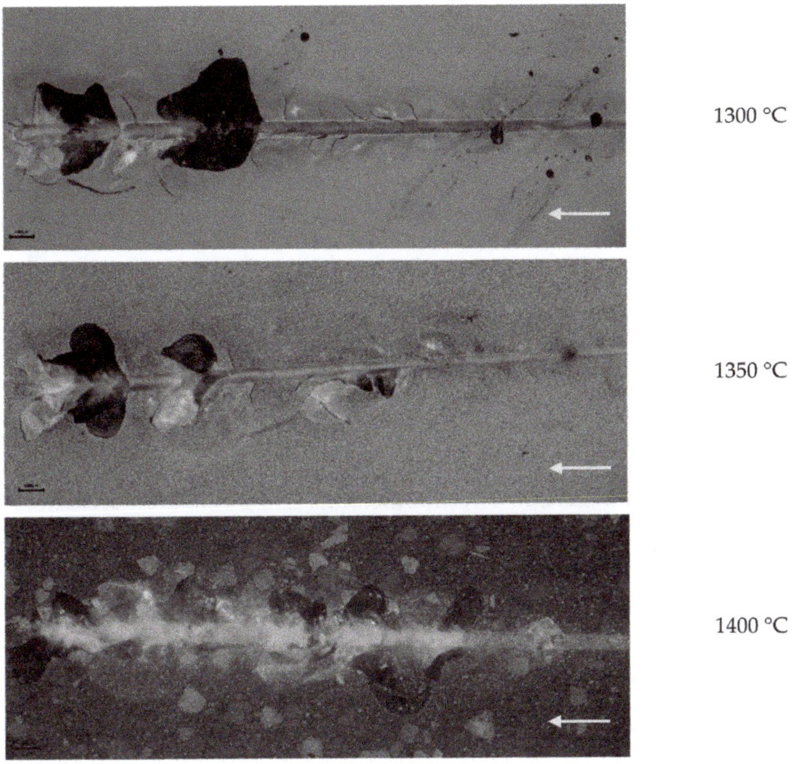

Figure 15. Microscopic observations of the parts of a scratch made on different transparent spinel (arrow: scratching direction).

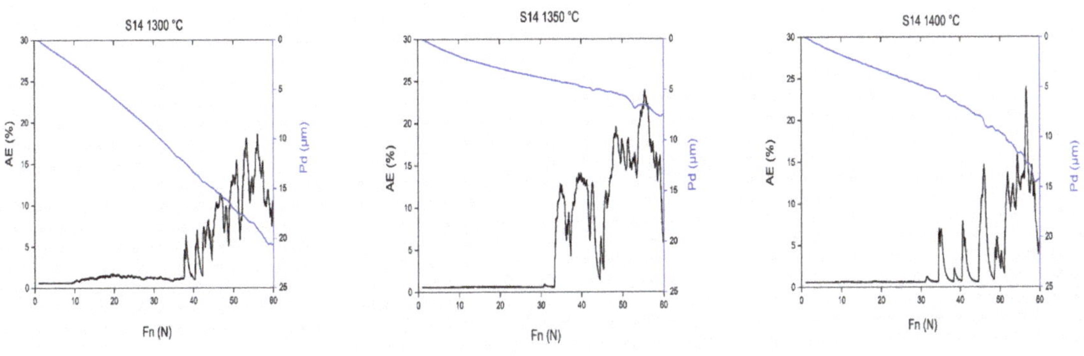

Figure 16. Recorded acoustic emission activities and corresponding penetration depth as a function of applied normal load for samples sintered at different temperatures.

4. Conclusions

The present study demonstrates the significant impact of SPS sintering temperature on the mechanical and optical properties of transparent spinel. High sintering temperatures lead to an increase in carbon contamination and promote the coalescence of residual pores, leading to the degradation in the optical properties. Conversely, it was observed that the best mechanical properties were achieved at low sintering temperatures, attributable to the presence of a small average grain size.

The main conclusions from this study are summarized as the following:

– The use of commercial nano-powder of spinel S25CRX 14, having a high chemical and crystallographic purity, results in dense (>99.9) and nanostructured-sintered pellets. It was also revealed that the optimal optical properties values found for samples sintered at 1350 °C, at a wavelength of 550 nm, are: (1) a real in-line transmission of 72.3%, (2) a total reflection RT of 13.1%, and (3) an absorbed light intensity of 23.5%.

– The samples sintered at 1350 °C led to the best compromise for mechanical properties, in particular, Vickers hardness (Hv ≈ 16 GPa), fracture toughness ($K_{IC} \approx 1.4$ MPa $\sqrt{m}$), and elastic modulus (E ≈ 272 GPa).

– The tribological study of the transparent spinel found that the spinel sintered at 1350 °C exhibited relatively high wear resistance with small track width and a low friction coefficient (0.35).

The results of this study also indicate that transparent ceramics were obtained with optimum transparency for sintering at 1350 °C. For this sintering temperature, a good combination of the mechanical properties was obtained. This transparent ceramic can be exploited in a wide range of applications, optical lenses and transparent armor.

Author Contributions: Conceptualization, K.H., S.B. (Salim Benaissa), A.C., S.B. (Sofiane Bouheroum), A.A., M.H. and G.F.; methodology, K.H., S.B. (Salim Benaissa), S.B. (Sofiane Bouheroum), M.H. and G.F.; validation, S.B. (Salim Benaissa), M.H. and G.F.; formal analysis, S.B. (Salim Benaissa), M.H. and G.F.; writing—original draft preparation, K.H., S.B. (Salim Benaissa), A.C., S.B. (Sofiane Bouheroum), M.H. and G.F.; writing—review and editing, K.H., S.B. (Salim Benaissa), M.H. and G.F.; visualization, K.H., S.B. (Salim Benaissa), M.H. and G.F.; supervision, K.H., S.B. (Salim Benaissa), M.H. and G.F.; All authors have read and agreed to the published version of the manuscript.

Funding: This research received no external funding.

Institutional Review Board Statement: Not applicable.

Informed Consent Statement: Not applicable.

Data Availability Statement: Data available upon request.

Conflicts of Interest: The authors declare no conflict of interest.

References

1. Johnson, R.; Biswas, P.; Ramavath, P.; Kumar, R.S.; Padmanabham, G. Transparent polycrystalline ceramics: An overview. *Trans. Indian. Ceram. Soc.* **2012**, *71*, 73–85. [CrossRef]
2. Shi, Z.; Zhao, Q.; Guo, B.; Ji, T.; Wang, H. A review on processing Polycrystalline magnesium aluminate spinel ($MgAl_2O_4$): Sintering techniques, material properties and machinability. *Mater. Des.* **2020**, *193*, 108858. [CrossRef]
3. Zegadi, A.; Kolli, M.; Hamidouche, M.; Fantozzi, G. Transparent $MgAl_2O_4$ spinel fabricated by spark plasma sintering from commercial powders. *Ceram. Int.* **2018**, *44*, 18828–18835. [CrossRef]
4. Kim, B.N.; Dash, A.; Kim, Y.; Morita, K.; Yoshida, H.; Li, J.G.; Sakka, Y. Low-temperature spark plasma sintering of alumina by using SiC molding set. *J. Ceram. Soc. Jpn.* **2016**, *124*, 1141–1145. [CrossRef]
5. Gajdowski, C.; D'Elia, R.; Faderl, N.; Böhmler, J.; Lorgouilloux, Y.; Lemonnier, S.; Leriche, A. Mechanical and optical properties of $MgAl_2O_4$ ceramics and ballistic efficiency of spinel based armour. *Ceram. Int.* **2022**, *48*, 18199–18211. [CrossRef]
6. Dudina, D.V.; Bokhonov, B.B.; Olevsky, E.A. Fabrication of porous materials by spark plasma sintering: A review. *Materials* **2019**, *12*, 541. [CrossRef]
7. Sokol, M.; Ratzker, B.; Kalabukhov, S.; Dariel, M.P.; Galun, E.; Frage, N. Transparent Polycrystalline Magnesium Aluminate Spinel Fabricated by Spark Plasma Sintering. *Adv. Mater.* **2018**, *30*, 1706283. [CrossRef]
8. Morita, K.; Kim, B.N.; Yoshida, H.; Keijiro, H.; Sakka, Y. Distribution of Carbon Contamination in $MgAl_2O_4$ Spinel Occurring during Spark-Plasma-Sintering (SPS) Processing: I—Effect of Heating Rate and Post-Annealing. *J. Eur. Ceram. Soc.* **2017**, *38*, 2588–2595. [CrossRef]
9. Sokol, M.; Kalabukhov, S.; Dariel, M.P.; Frage, N. High-pressure spark plasma sintering (SPS) of transparent Polycrystalline Magnesium Aluminate Spinel (PMAS). *J. Eur. Ceram. Soc.* **2014**, *34*, 4305–4310. [CrossRef]
10. Sokol, M.; Halabi, M.; Mordekovitz, Y.; Kalabukhov, S.; Hayun, S.; Frage, N. An inverse Hall-Petch relation in nanocrystalline $MgAl_2O_4$ spinel consolidated by high pressure spark plasma sintering (HPSPS). *Scr. Mater.* **2017**, *139*, 159–161. [CrossRef]
11. Akinribide, O.J.; Mekgwe, G.N.; Akinwamide, S.O.; Gamaoun, F.; Abeykoon, C.; Johnson, O.T.; Olubambi, P.A. A review on optical properties and application of transparent ceramics. *J. Mater. Res. Technol.* **2022**, *21*, 712–738. [CrossRef]

12. Liu, L.; Morita, K. Effect of sintering conditions on optical and mechanical properties of $MgAl_2O_4/Al_2O_3$ laminated transparent composite fabricated by spark-plasma-sintering (SPS) processing. *J. Eur. Ceram. Soc.* **2022**, *42*, 2487–2495. [CrossRef]
13. Nečina, V.; Hostaša, J.; Pabst, W.; Veselý, M. Magnesium fluoride (MgF2)—A novel sintering additive for the preparation of transparent YAG ceramics via SPS. *J. Eur. Ceram. Soc.* **2022**, *42*, 3290–3296. [CrossRef]
14. Chen, X.; Wu, Y. Fabrication and optical properties of highly transparent MgO ceramics by spark plasma sintering. *Scr. Mater.* **2019**, *162*, 14–17. [CrossRef]
15. Li, J.G.; Ikegami, T.; Lee, J.H.; Mori, T. Fabrication of translucent magnesium aluminum spinel ceramics. *J. Am. Ceram. Soc.* **2000**, *83*, 2866–2868. [CrossRef]
16. Ganesh, I. A review on magnesium aluminate ($MgAl_2O_4$) spinel: Synthesis, processing and applications. *I. Mater. Rev.* **2013**, *58*, 63–112. [CrossRef]
17. Wang, X.; Fang, M.; Zhang, L.C.; Ding, H.; Liu, Y.G.; Huang, Z.; Huang, S.; Yang, J. Solid particle erosion of alumina ceramics at elevated temperature. *Mater. Chem. Phys.* **2013**, *139*, 765–769. [CrossRef]
18. Bousbaa, C.; Madjoubi, A.; Hamidouche, M.; Bouaouadja, N. Effect of annealing and chemical strengthening on soda lime glass erosion wear by sand blasting. *J. Eur. Ceram. Soc.* **2003**, *23*, 331–343. [CrossRef]
19. Murugesh, L.; Scattergoog, R.O. Effect of erodent properties on the erosion of alumina. *J. Mater. Sci.* **1991**, *26*, 5456–5466. [CrossRef]
20. Oka, Y.I.; Mihara, S.; Yoshida, T. Impact-angle dependence and estimation of erosion damage to ceramic materials caused by solid particle impact. *Wear* **2009**, *267*, 129–135. [CrossRef]
21. Lallemant, L.; Garnier, V.; Bonnefont, G.; Marouani, A.; Fantozzi, G.; Bouaouadja, N. Effect of solid particle impact on light transmission of transparent ceramics: Role of the microstructure. *Opt. Mater.* **2014**, *37*, 352–357. [CrossRef]
22. Von Helden, S.; Malzbender, J.; Krüger, M. Mechanical properties, wear resistance and surface damage of glasses and $MgAl_2O_4$ spinel ceramic after abrasion and scratch exposure. *Ceram. Int.* **2019**, *45*, 10765–10775. [CrossRef]
23. Zhou, J.; Bahadur, S. The effect of material composition and operational variables on the erosion of alumina ceramics. *Wear* **1991**, *150*, 343–354. [CrossRef]
24. Ćurković, L.; Kumić, I.; Grilec, K. Solid particle erosion behaviour of high purity alumina ceramics. *Ceram. Int.* **2011**, *37*, 29–35. [CrossRef]
25. Morita, K.; Kim, B.N.; Hiraga, K.; Yoshida, H. Spark-plasma-sintering condition optimization for providing transparent $MgAl_2O_4$ spinel polycristal. *J. Am. Ceram. Soc.* **2009**, *92*, 1208–1216. [CrossRef]
26. Bonnefont, G.; Fantozzi, G.; Trombert, S.; Bonneau, L. Fine-grained transparent $MgAl_2O_4$ spinel obtained by spark plasma sintering of commercially available nanopowders. *Ceram. Int.* **2012**, *38*, 131–140. [CrossRef]
27. Benaissa, S.; Hamidouche, M.; Kolli, M.; Bonnefont, G.; Fantozzi, G. Characterization of nanostructured $MgAl_2O_4$ ceramics fabricated by spark plasma sintering. *Ceram. Int.* **2016**, *42*, 8839–8846. [CrossRef]
28. Apetz, R.; Van Bruggen, M.P.B. Transparent Alumina: A Light-Scattering Model. *J. Am. Ceram. Soc.* **2003**, *86*, 480–486. [CrossRef]
29. *ASTM C1327-03*; Standard Test Method for Vickers Indentation Hardness of Advanced Ceramics. ASTM International: West Conshohocken, PA, USA, 2003.
30. Benaissa, S.; Hamidouche, M.; Kolli, M.; Fantozzi, G. Optical and mechanical Characterization of transparent α-alumina fabricated by Spark Plasma Sintering. *Int. J. Appl. Ceram. Technol.* **2019**, *16*, 638–646. [CrossRef]
31. Anstis, R.G.; Chantikul, P.; Lawn, B.R.; Marshall, D.B. A critical evaluation of indentation techniques for measuring fracture toughness: I, direct crack measurements. *J. Am. Ceram. Soc.* **1981**, *64*, 533–538. [CrossRef]
32. Kulyk, V.; Duriagina, Z.; Vasyliv, B.; Vavrukh, V.; Kovbasiuk, T.; Lyutyy, P.; Vira, V. The Effect of Sintering Temperature on the Phase Composition, Microstructure, and Mechanical Properties of Yttria-Stabilized Zirconia. *Materials* **2022**, *15*, 2707. [CrossRef]
33. Malou, Z.; Hamidouche, M.; Bouaouadja, N.; Fantozzi, G. Thermal shock resistance of sodalime glass eroded by sandblasting. *Glass Struct. Eng.* **2019**, *5*, 147–153. [CrossRef]
34. Razavi, R.S.; Ahsanzadeh-Vadeqani, M.; Barekat, M.; Naderi, M.; Hashemi, S.H.; Mishra, A.K. Effect of sintering temperature on microstructural and optical properties of transparent yttria ceramics fabricated by spark plasma sintering. *Ceram. Int.* **2016**, *42*, 7819–7823. [CrossRef]
35. Hammoud, H.; Garnier, V.; Fantozzi, G.; Lachaud, E.; Tadier, S. Mechanism of carbon contamination in transparent $MgAl_2O_4$ and $Y_3Al_5O_{12}$ ceramics sintered by spark plasma sintering. *Ceramics* **2019**, *2*, 48. [CrossRef]
36. Rothman, A.; Kalabukhov, S.; Sverdlov, N.; Dariel, M.P.; Frage, N. The Effect of Grain Size on the Mechanical and Optical Properties of Spark Plasma Sintering-Processed Magnesium Aluminate Spinel $MgAl_2O_4$. *Int. J. Appl. Ceram. Technol.* **2012**, *11*, 146–153. [CrossRef]
37. Morita, K.; Kim, B.N.; Yoshida, H.; Keijiro, H.; Sakka, Y. Assessment of carbon contamination in $MgAl_2O_4$ spinel during spark-plasma-sintering (SPS) processing. *J. Ceram. Soc. Jpn.* **2015**, *123*, 983–988. [CrossRef]
38. Bernard-Granger, G.; Benameur, N.; Guizard, C.; Nygren, M. Influence of graphite contamination on the optical properties of transparent spinel obtained by spark plasma sintering. *Scr. Mater.* **2009**, *60*, 164–167. [CrossRef]
39. Morita, K.; Kim, B.N.; Yoshida, H.; Sakka, Y. Distribution of carbon contamination in oxide ceramics occurring during spark-plasma-sintering (SPS) processing: II-Effect of SPS and loading temperatures. *J. Eur. Ceram. Soc.* **2018**, *38*, 2596–2604. [CrossRef]
40. Morita, K.; Kim, B.N.; Yoshida, H.; Keijiro, H.; Sakka, Y. Spectroscopic study of the discoloration of transparent $MgAl_2O_4$ spinel fabricated by spark-plasma-sintering (SPS) processing. *Acta Mater.* **2015**, *84*, 9–19. [CrossRef]

41. Goldstein, A.; Goldenberg, A.; Yeshurun, Y.; Hefetz, M. Transparent $MgAl_2O_4$ spinel from a powder prepared by flame spray pyrolysis. *J. Am. Ceram. Soc.* **2008**, *91*, 4141–4144. [CrossRef]
42. Krell, A.; Bales, A. Grain Size-Dependent Hardness of Transparent Magnesium Aluminate Spinel. *Int. J. Appl. Ceram. Technol.* **2011**, *8*, 1108–1114. [CrossRef]
43. Tokariev, O.; Steinbrech, R.W.; Schnetter, L.; Malzbender, J. Micro-and macro-mechanical testing of transparent $MgAl_2O_4$ spinel. *J. Mater. Sci.* **2012**, *47*, 4821–4826. [CrossRef]
44. Hamidouche, M.; Khennafi-Benghalem, N.; Jacommet, S.; Fides, M.; Hvizdoš, P. Study of Ni63-Co37 nanocrystalline electrode-posited alloy as anti-wear coating on mild steel substrate. *Mater. Res. Express* **2019**, *6*, 126557. [CrossRef]
45. Mroz, T.; Goldman, L.M.; Gledhill, A.D.; Li, D.; Padture, N.P. Nanostructured, infrared-transparent magnesium-aluminate spinel with superior mechanical properties. *Int. J. Appl. Ceram. Technol.* **2012**, *9*, 83–90. [CrossRef]
46. Le Houerou, V.; Sangleboeuf, J.C.; Derianou, S.; Rouxel, T.; Duisit, G. Surface damage of soda–lime–silica glasses: Indentation scratch behavior. *J. Non-Cryst. Solids* **2003**, *316*, 54–63. [CrossRef]
47. Huang, L.; Yao, W.; Mukherjee, A.K.; Scoenung, J.M. Improved mechanical behaviour and plastic deformation capability of ultrafine grain alumina ceramics. *J. Am. Ceram. Soc.* **2012**, *95*, 379–385. [CrossRef]
48. Huang, L.; Zhang, Z.; Zhao, Y.; Yao, W.; Mukherjee, A.K.; Scoenung, J.M. Scratch-induced deformation in fine- and ultrafine-grained bulk alumina. *Scr. Mater.* **2010**, *63*, 528–531. [CrossRef]
49. Klecka, M.; Subhash, G. Grain size dependence of scratch-induced damage in alumina ceramics. *Wear* **2008**, *265*, 612–619. [CrossRef]

ceramics

MDPI

Review

Speciation of 3d Elements in Spinel Versus Corundum: Elucidating the Interplay Between Ligand Field, Structural Dissimilarities and Processing Conditions

Adrian Goldstein [1] and Alessio Zandonà [2,*]

1 Israel Ceramic and Silicate Institute, Technion City, Haifa 3200003, Israel
2 Institute of Non-Metallic Materials, Clausthal University of Technology, 38678 Clausthal-Zellerfeld, Germany
* Correspondence: alessio.zandona@tu-clausthal.de

Abstract: The simultaneous analysis of optical and electronic paramagnetic resonance spectra of all 3d metals, doped into transparent α-Al_2O_3 and $MgAl_2O_4$ spinel, was effectuated with a view of establishing the speciation pattern of the dopants. The examination of these patterns enabled the revelation of certain regularities (rules) affecting the correlation between the physical factors controlling the process and speciation patterns. It was observed that structural dissimilarities between the lattices significantly affected the correlation. Thus, the spinel lattice was found to impose the accommodation of the dopants as 2+ cations replacing native Mg^{2+} ions located in tetrahedral sites, with the process concerning only the late 3d elements. The difference in behavior between the early and late 3d elements is mostly caused by the increase in ionization potential along the series. In alumina, the dopants are accommodated as 3+ cations in octahedral sites; 6-coordinated 2+ cation stabilization is feasible but requires extremely reductive conditions for late 3d elements.

Keywords: $MgAl_2O_4$ spinel; corundum; transition metal speciation; optical spectroscopy

1. Introduction

1.1. Background

Transition metal cations (TM+) entering transparent solid hosts (TSHs) as a result of doping with transition metals (TMs) determine, among other things, their overall response in terms of optical absorption, luminescence, and electron paramagnetic resonance. Important applications in domains like solid-state lasers, Q switches, phosphors, scintillators, pigments, or optical filters depend on these profiles; in other cases, TM impurities hamper applications [1].

The scientific background used for the further development of these applications includes, as an essential part, the results of various spectroscopy techniques applied to TSHs doped with TM+. Therefore, it is important to measure, explain, and predict the profile of such spectra. For the last two tasks (especially the latter), it is crucial to identify how TMs might speciate when doped into TSHs, for instance with the aid of variants of the ligand field theory (LFT); two examples of good textbooks are [2,3].

In order to understand speciation, it is essential to know the way various physical factors influence the process. Previous scientific work indicates that a speciation pattern is, in detail, the result of the interaction of quite a large number of factors. However, in many cases, only a few of these exert a dominant role, as described in the following.

https://doi.org/10.3390/ceramics8010016

The first factor is the ionization energy (IE). Going over the 3d series from left to right, two segments exhibiting continuous linear increase can be identified, separated by a drop (from Mn to Fe, followed by a recovery at Co). The energy difference between the group defined by the first segment (Ti, V, and Fe) and the second one (Mn, Co, Ni, and Cu) is remarkable, exceeding 6 eV. Below, we label the Ti, V, and Cr as "early elements" and the rest as "late elements" in the series.

The second factor is Δr, i.e., the difference between the radius of a native cation and that of a TM+ seeking to replace it. The probability of such a replacement is also influenced by the difference between the actual site shape and the one that would be optimal for an incoming cation. Indeed, any TM+ entering a host invariably tends to remodel the occupied site towards a preferred symmetry. Thus, it would be more accurate to consider a "geometric factor", rather than only Δr. Since ionic radii *r* are repeatedly discussed below, those of the native cations are recalled here as reported in the literature in the case of lattices possessing exclusively ionic bonding: for $^{VI}Mg^{2+}$, $r = 72$ pm; for $^{IV}Mg^{2+}$, $r = 57$ pm; for $^{VI}Al^{3+}$, $r = 54$ pm; for $^{IV}Al^{3+}$, $r = 39$ pm.

The third factor needing consideration is a potential stabilization mechanism due to the geometry of a generated complex, which is specific to the electronic configuration of the TM+ generating the complex. The number of possible mechanisms is small, such as the crystal field stabilization energy (CFSE); the stabilization conferred by a few special levels of "3d" sub-shell filling, viz., d^0 (empty), d^5 (half complete), d^{10} (full); and Jahn–Teller coupling (J–T).

The fourth stabilization factor revolves around the fact that TM+ incorporation by a host requires charge compensation. If energetically reasonable mechanisms are not available for this, the considered TM+ cannot be doped into the host, albeit fitting from other points of view. Therefore, it is important to pinpoint the available charge compensation options. Unfortunately, it is rarely possible to figure out all the options and rank them. The most elusive are clusters of intrinsic and extrinsic point defects. These may form complex structures, acting as low-energy compensators. Even when the analysis is performed post-factum, it is difficult to accurately determine the actual compensation mechanism from the examination of the spectra. An example of a controllable mechanism is the stabilization of low TM+ oxidation states by using oxygen-scarce atmospheres during specimen fabrication (see Section 3).

The fifth factor is the Madelung electrostatic energy of the lattice. Importantly, this energy is lower if doping is performed by substituting a native cation and not via targeting interstitial vacant sites.

Also important is the redox state of the atmosphere at the high temperatures that most specimens attain during fabrication. Studies on glass especially manifest how powerful the effect of this factor may be [4].

Last but not least, one has to note the decisive influence of a seventh factor: the real amount of negative charge (not the formal 2−) carried by the oxide ligands of a given host. Below, it will be labeled "δ−" (viz., the ligands will be denoted as $O^{\delta-}$) and called "basicity", following the reasoning outlined by J.A. Duffy for glasses, which is also valid in this context. Calculations like the ones made for glass are not feasible here. However, a comparison of OAS strongly suggests that the average basicity is quite similar in the two hosts relevant to this work, with corundum exhibiting a slightly higher basicity.

These factors represent the influence of three basic aspects: electronic configuration of the TM+ into which a given TM may speciate, host structure, and fabrication conditions (especially redox properties of the atmosphere and thermal treatment profile).

1.2. Research Objectives

As suggested in the Introduction, this research aimed at a better understanding of TM dopants in transparent solids. The first goal was the accurate determination of the speciation pattern of all 3d TMs in α-Al_2O_3 and $MgAl_2O_4$, as this information was previously not available.

The second goal was to elucidate the speciation of TM+ evaluating interplay between the stabilization factors listed in Section 1.1 and the change in the host structure from α-Al_2O_3 to $MgAl_2O_4$. These particular hosts were selected because the corundum lattice offers only filled crystallographic sites with an octahedral symmetry, while the spinel structure includes both octahedral and tetrahedral sites. In addition, only Al^{3+} cations are found in α-Al_2O_3, while both Al^{3+} and Mg^{2+} are present in $MgAl_2O_4$. Both hosts are relevant for optical applications. For example, Cr-Ti or Ti-Fe co-doped α-Al_2O_3 finds use as a lasing material, as well as gemstone. Similarly, spinel crystals doped with TM+ represent attractive gemstones, while Co^{2+} doping yields efficient passive absorber-type Q-switches.

1.3. Research Approach

Available spectral data (published and recorded in-house) referring to all 3d TMs doped into α-Al_2O_3 and $MgAl_2O_4$ were collected and analyzed. More specifically, results of optical absorption (OAS), luminescence (LUMI), and electron paramagnetic resonance (EPR) spectra were used. Conclusions were then drawn based on a comparison of the results.

1.4. The Corundum and Spinel Lattices

First, a few geometrical characteristics of the two lattices under consideration are recalled in order to facilitate spectral data interpretation without the need to consult additional sources. Spinel and corundum are used in the following interchangeably as names for the chemical compounds α-Al_2O_3 and $MgAl_2O_4$.

1.4.1. Corundum (α-Al_2O_3)

The oxide ions of the lattice (formally O^{2-}) form a hexagonal close packed (hcp) arrangement, stacked in an A-B-A-B sequence. Such an arrangement has octahedral and tetrahedral voids: a 2/3 fraction of the octahedral voids is occupied by Al^{3+} cations, while no tetrahedral voids are occupied. The lattice can be described using both a hexagonal and a trigonal crystallographic reference system and the point group is D_{3d} (Schönflies notation). The local rotational symmetry of the Al^{3+} site is usually defined as C_3 or C_{3v}, depending on the literature source, and the approximation is good enough for ligand field calculations; in truth, very slight non-axial distortions are frequently present. In the case of the smallest convenient trigonal unit cell (two formula units), the a and c parameters equal, respectively, 4.758 Å and 12.922 Å (the angles are $\alpha = \beta = 90°$ and $\gamma = 120°$).

The six oxide ions that sit closest to Al^{3+} define the vertices of a trigonally distorted octahedron [5]. Two Al nearest neighbors share the triangle of oxygen atoms lying between them along the c axis, with the Al-O distance being 1.969 Å (Figure 1) [6]. For each Al^{3+} ion, the second triangle defined by oxygen ions is larger and closer to the central Al atom, with an Al-O distance equal to 1.856 Å. Considering a more accurate symmetry description, the oxygen triangles lying between two Al^{3+} cations are actually slightly rotated away (by ~3°) from the σ_v planes, so that symmetry is further reduced (some authors report D_{3d} symmetry as being more accurate [6]). As for cation interaction (an aspect influencing sometimes spectral profiles, e.g., in Fe-Ti dopant pairs), it is worth noting that part of the oxygen octahedra shares edges, whereas others are face-sharing in directions parallel and perpendicular, respectively, to the c axis.

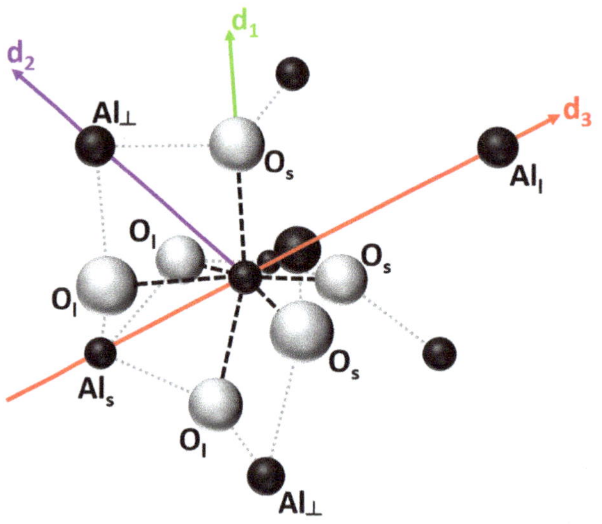

Figure 1. View of the local structure of α-Al_2O_3, showing ionic positions around a reference Al^{3+} cation in the lattice; the symmetry of the polyhedron formed by the oxide ions around Al^{3+} is a slightly distorted octahedron (C_3 symmetry). The O_s oxide ions are at a distance R_s = 1.857 Å, while O_l ions lie at R_l = 1.969 Å. Along the C_3 axis ($c = d_3$), one of the cationic neighbors lies at 2.65 Å (Al_s) and the other lies at 3.80 Å (Al_l). Figure plotted based on the structural data presented in [6].

An important parameter is w, which represents the distance between an Al^{3+} cation and the point lying midway (along the c axis) between two other Al^{3+} ions. When a TM+ replaces a native cation, the Δr will modify distances and possibly angles within the crystallographic site, i.e., the w parameter will be modified [5]. There are also other factors affecting the geometry of a given site following cationic replacement [7]; for instance, the incoming TM+ may be subject to a significant J-T effect [8].

1.4.2. Spinel ($MgAl_2O_4$)

The cubic (O_h point group) lattice type called "spinel" is found in AB_2X_4 compounds, where X may also be the oxide anion and the A and B cations may assume various oxidation states between 2+ and 4+. The archetype of the spinel-type lattice is the double oxide $MgAl_2O_4$ (a mineral also called "spinel"). In the cubic lattice of the material (Figure 2), the oxide ions take a face-centered cubic (fcc) arrangement. A convenient unit cell for lattice geometry description includes eight unit formulas, with a lattice parameter a = 8.0898 Å. In addition, an important parameter is u, i.e., the distance between the oxide ions. In contrast with expectations based on cations' radii, the octahedral sites accommodate Al^{3+} and the tetrahedral ones accommodate Mg^{2+}.

The Mg^{2+} cations occupy 8 of the 64 available tetrahedral sites (the local symmetry is T_d) and the Al^{3+} cations fill 16 of the 32 octahedral sites (the local symmetry is O_h) within a unit cell. In actual spinel crystals (including $MgAl_2O_4$), the oxide ions are slightly displaced from the ideal position along the [111] direction (u is usually slightly higher), so that the O_h symmetry around Al^{3+} is slightly distorted [9]. Thus, the volume of the tetrahedral sites increases and that of the octahedral ones correspondingly decreases. This process may be accentuated when the native cations are replaced with TM+. There is no contact among the tetrahedra in the lattice, while the octahedra succeed in establishing edge contact in certain directions. Various values have been reported in the literature for the interionic distances, such as 1.855 Å for the Al-O pair and 1.967 Å for the Mg-O pair [10].

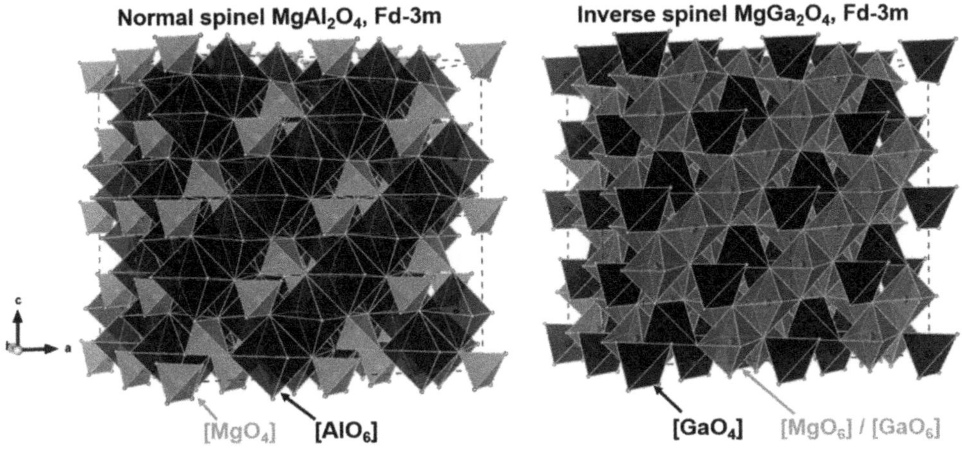

(a) (b)

Figure 2. Normal and inverse spinel structures, in their most archetypal occurrences: (**a**) $MgAl_2O_4$ and (**b**) $MgGa_2O_4$.

A fairly specific feature of this lattice is the tendency towards inversion, justified in terms of thermodynamic considerations [11]. This phenomenon is quantified by the inversion factor *i*, which varies between 0 and 1. Therefore, a more accurate description of a spinel lattice site occupancy is seen in Equation (1):

$$[^{VI}Al^{3+}]^0 + [^{IV}Mg^{2+}]^0 \leftrightarrow [^{IV}Al^{3+}]^+ + [^{VI}Mg^{2+}]^- \quad (1)$$

wherein roman numerals represent the coordination numbers of the cations, while residual charges on the respective polyhedra are reported outside square brackets. Depending on the value of *i*, a more or less large fraction of Mg^{2+} cations fills octahedral sites, while a similar amount of Al^{3+} sits inside tetrahedral ones. The locations occupied by 4-coordinated Al^{3+} and 6-coordinated Mg^{2+} are called antisites. Vis à vis the lattice, these are electrically charged defects; their size and distortion level differ from those of normal sites and they often interact forming dipolar units. A normal spinel has $i = 0$ and a fully inverse [11] one has $i = 1$. In practice, such *i* values are rarely seen. Some of the natural $MgAl_2O_4$ crystals have *i* values very close to zero, which rise to ~0.3 on heating. Artificial crystals synthesized at a high temperature usually involve a certain degree of inversion. In Figure 2, a normal and inverse lattice are compared. The fraction of antisites may vary depending on aspects like the exact stoichiometry of the specimen and the synthesis route.

1.4.3. Structural Comparison

A comparison of the lattices shows that the most conspicuous structural differences between them are with regard to the geometry of the sites offered to incoming TM+. Both lattices offer virtually cubic octahedral sites that are occupied by trivalent native cations, while the spinel structure also entails tetrahedral sites filled by divalent cations. However, a more accurate description reveals that the octahedral sites in corundum are somewhat more distorted (towards a trigonal or even lower symmetry) than in spinel (exhibiting a slighter tetragonal distortion). Moreover, the Al^{3+}-O^{2-} distances are larger in corundum. From the total number of available sites, a larger fraction is occupied by Al^{3+} in corundum. The octahedra share edges along certain directions in spinel, while edge-sharing is accompanied by face-sharing in corundum. In spinel, the cation-to-oxide distance is larger in A-type sites ($^{IV}Mg^{2+}$ sites) than in B-type sites ($^{VI}Al^{3+}$ sites). In spinel, a strain caused by doping a TM+ with a certain Δr leads to an increase in u, enlarging the tetrahedra without causing distortion; simultaneously, it causes a reduction in the volume of the octahedral sites and their further distortion.

When TM+ cations are introduced into corundum and spinel hosts, further TM+-specific site distortion usually occurs, owing to the difference in the geometry of the native site and that of the complex formed by the entering TM+. The shape of the complex may be affected in certain cases by the so-called J-T coupling, whose identification is, however, not so simple. The effect is active only if the starting complex is perfectly cubic and an adequate symmetry of the electronic and vibrational states of the lattice exists.

2. Spectral Data and Speciation Pattern Analysis

The speciation pattern of TMs may be established by analyzing various types of spectra. Some common examples in addition to those used in this work are Raman spectroscopy, X-ray photo-electron spectroscopy (XPS), Auger spectroscopy (AES), extended X-ray absorption fine structure (EXAFS), and the related X-ray absorption near-edge spectroscopy (XANES). No method is free from errors and, ideally, data from multiple sources should be compared in order to improve the reliability of conclusions. The methods we selected (OAS, LUMI, and EPR) are among the most informative, with extensive literature coverage.

Spectral data were derived from both single crystals and ceramic (polycrystalline) specimens. Both types of specimens usually carry a certain amount of fabrication-related defects such as pores, undesired secondary phases, random stress fields, impurity atoms, and intrinsic point defects, although recent synthesis technologies can yield samples of very high optical quality. In many cases, ceramic specimens can attain comparable light transmission, homogeneity of dopant distribution, and stress fields as the best single crystals. However, the spectral features discussed here can be observed even in specimens of lesser quality, especially relying on the higher detail of measurements performed under cryogenic conditions. Single crystals have the significant advantage of enabling measurements along selected crystallographic directions, as this may be critical (especially in EPR) to obtain unambiguous results.

While direct presentation of all discussed spectra in graphic form would be helpful to the reader, the amount of figures required for such an approach would be prohibitively large. Therefore, only the most pertinent data are shown here, while relevant tabulated data are collected in Tables 1 and 2 in terms of band positions. Table 1 includes the early TMs, whose common feature is the lack of a tendency toward assuming 2+ states in spinel. Table 2 provides information on the late TMs, which instead exhibit such a tendency markedly.

Table 1. Spectroscopic data (OAS for optical absorption, EPR for electron paramagnetic resonance, LUMI for luminescence spectroscopy) of Ti^{3+}, V^{3+}, and Cr^{3+} in α-Al_2O_3 and $MgAl_2O_4$ hosts. Abbreviations and symbols: "Oct" for octahedral coordination, "Tet" for tetrahedral coordination, "sc" for single crystal, "sh" for shoulder, values in parentheses for weak signals.

TM^+	**OAS (cm^{-1})**		**EPR**		**LUMI (cm^{-1})**	
	Corundum	**Spinel**	**Corundum**	**Spinel**	**Corundum**	**Spinel**
Ti^{3+} (d^1)	(11,800) 18,000 sh 20,500	13,200 (19,600)	$g_{II} = 1.07$ $g_{\perp} < 0.1$	$g_{eff} \approx 1.95$	13,300	-
	Oct	Tet/Oct	Oct	Oct	Oct	-
	[5] sc	[12] sc	[13] sc	[14] sc	[15] sc	-

Table 1. *Cont.*

TM^+	OAS (cm^{-1})		EPR		LUMI (cm^{-1})	
	Corundum	Spinel	Corundum	Spinel	Corundum	Spinel
V^{3+} (d^2)	18,000 25,000 (~33,000) Oct [5] sc	17,200 25,200 Oct [12] sc	- - -	- - -	- - -	- - -
Cr^{3+} (d^3)	(14,400) 17,900 25,000 Oct [5] sc	18,000 25,200 37,000 Oct [12] sc	$g \approx 1.98$ $g = 4.0$ Oct [16] sc	$g = 1.96$ $g = 3.82$ Oct [17] sc	14,400 (R-line) 14,534 (R-line) 14,771 (N-line) 14,085 (side band) Oct [18] sc	14,620 (R_1) 13,700–14,900 (range) Oct [19] sc

Table 2. Spectroscopic data (OAS for optical absorption, EPR for electron paramagnetic resonance, LUMI for luminescence spectroscopy) of Mn^{2+}, Fe^{2+}, Co^{2+}, Ni^{2+}, and Cu^{2+} in α-Al_2O_3 and $MgAl_2O_4$ hosts. Abbreviations and symbols: "Oct" for octahedral coordination, "Tet" for tetrahedral coordination, "sc" for single crystal, "pow" for powder, "*" for calculated values, "^{m}" for Mn^{2+}-related signals (partially) masked by the stronger signatures of coexisting Mn^{3+}, parentheses for weak signals.

TM^+	OAS (cm^{-1})		EPR		LUMI (cm^{-1})	
	Corundum	Spinel	Corundum		Corundum	Spinel
Mn^{2+} (d^5)	25,000 ^{m} 27,500 ^{m} 29,000 Oct [20] sc	20,000 21,800 22,700 26,300 29,400 Tet [21] sc	$S = 5/2$ $I = 5/2$ $g_{\parallel} = 2.0017$ $g_{\perp} = 2.000$ Oct [22] sc	$g \approx 2$ Tet/Oct [23] sc	~15,000 Oct [20] sc	16,000 19,400 Tet/Oct [21] sc
Fe^{2+} (d^6)	11,200 Oct [24] sc	4500 Tet [12] sc	- - -	- - -	- - -	- - -
Co^{2+} (d^7)	9100 21,400 22,800 Oct [25] sc	~4000 * 6500 16,700 Tet [26] cer	- - -	- - -	- - -	15,400 (11,100) Tet [26] cer
Ni^{2+} (d^8)	~10,000 * 18,050 27,500 Oct [25] sc	4500 9500 13,500 23,000 27,400 Tet/Oct [12] sc	$g_{\parallel} = 2.196$ $g_{\perp} = 2.186$ Oct [27] sc	- - -	- - -	8100 Oct [28] sc
Cu^{2+} (d^9)	~9500 11,000 Oct [29] cer pow	6540 11,600 Tet/Oct [30] sc cer	$g_{\parallel} = 2.325$ $g_{\perp} = 2.048$ Oct [31] cer pow	-	7800 Oct [29] cer pow	- - -

Finally, oxidation states and coordination numbers assumed by TM dopants in corundum and spinel hosts are summarized in Table 3 based on the interpretation of relevant spectral data (see below). The major (or unique) species occurring under usual fabrication conditions are given in bold; when detectable, minor species are shown within round parentheses, while species becoming stable only under special conditions are indicated in square brackets. Those species whose electronic configuration confers outstandingly high stability (by one mechanism or other) are underlined.

Table 3. Speciation of transition metal cations in the hosts analyzed within this work. Major species are in bold, while minor species are given in round brackets and species appearing only under special synthesis conditions appear in square brackets. Underlined species exhibit inherently high stability due to their electronic configuration. "?" labels species that, although not detected so far in the hosts under study, may be expected to occur in low amounts under suitable synthesis conditions.

Transition Metal	α-Al_2O_3	$MgAl_2O_4$ Spinel
	Species	**Species**
Ti	$^{VI}\mathbf{Ti^{3+}}$ $(\underline{^{VI}Ti^{4+}})$	$^{IV}\mathbf{Ti^{3+}}$, $^{VI}\mathbf{Ti^{3+}}$ $(\underline{^{VI}Ti^{4+}})$
V	$^{VI}\mathbf{V^{3+}}$ $[^{VI}V^{4+}]$	$^{VI}\mathbf{V^{3+}}$ $\underline{[^{VI}V^{5+}]}$
Cr	$\underline{^{VI}\mathbf{Cr^{3+}}}$	$\underline{^{VI}\mathbf{Cr^{3+}}}$
Mn	$^{VI}\mathbf{Mn^{3+}}$ $[\underline{^{VI}Mn^{2+}}, ^{VI}Mn^{4+}]$	$\underline{^{IV}\mathbf{Mn^{2+}}}$ $(^{VI}Mn^{2+})[^{VI}Mn^{3+}]$
Fe	$\underline{^{VI}\mathbf{Fe^{3+}}}$ $(^{VI}Fe^{2+})$	$^{IV}\mathbf{Fe^{2+}}$ $(^{VI}Fe^{2+}, \underline{^{IV}Fe^{3+}}, \underline{^{VI}Fe^{3+}})$
Co	$^{VI}\mathbf{Co^{3+}}$ $(^{VI}Co^{2+})[^{VI}Co^{4+}]$	$^{IV}\mathbf{Co^{2+}}$ $(^{VI}Co^{2+})$
Ni	$^{VI}\mathbf{Ni^{3+}}$ $[\underline{^{VI}Ni^{2+}}]$	$^{IV}\mathbf{Ni^{2+}}$ $(\underline{^{VI}Ni^{2+}})$
Cu	$\underline{^{VI}\mathbf{Cu^{3+}}}$ $[Cu^{2+}](Cu^{+}?)$	$^{IV}\mathbf{Cu^{2+}}$ $(^{VI}Cu^{2+}, Cu^{0}, Cu^{+}?)$

2.1. *Titanium*

2.1.1. α-Al_2O_3

Spectral analysis indicates that the host is able to accommodate $Ti^{4,3+}$. Concerning the low solubility of Ti^{4+}, relevant data are provided in [32]. Ti^{4+} (d^0) produces no d-d spectra [33], but it generates blue luminescence (~480nm) when located in octahedral sites [34], while no emission is recorded when it is located in tetrahedral sites. The OAS of the d^1 cation Ti^{3+} shows a strong band at 20,500 cm^{-1}, as well as a broad CT band and a shoulder at 18,000 cm^{-1} indicating distortion of the octahedral site [5,14,35]. Some authors specify the site symmetry as C_{3v} or lower [36], while others consider a tetragonal symmetry in view of J-T contribution [37]. The OAS spectrum of Ti^{3+} in corundum has its bands in the position shown in Table 1; a high-resolution figure appears in [38]. Table 1 shows the position of major bands in LUMI and EPR data [13,15,34,39,40], which confirm the existence of an axial site distortion.

2.1.2. $MgAl_2O_4$

Luminescence and OAS data [41–43] ascertain that some 6-coordinated Ti^{4+} is also accommodable in spinel crystals, similarly to what is seen in corundum. Under suitable conditions (i.e., strong oxidative atmosphere), the species may even be the only one accommodated in single crystals or polycrystalline ceramics. Suitable conditions for increasing Ti^{4+} content in single crystals are offered by the floating zone synthesis method, which can be used at low temperatures (1000 to 1200 °C) under atmospheres containing some oxygen. In Verneuil crystals, Ti^{4+} substitutes 6-coordinated Al^{3+} native cations after 24 h annealing at 1200 °C in air, oxidizing pre-existing Ti^{3+}. For crystals fabricated at temperatures around 1600 °C under oxidative or slightly reductive atmospheres, the spectral profile indicates that the dopant may completely remain as Ti^{4+}. If such specimens are re-heated under more reductive conditions, however, a small fraction of the dopant assumes a Ti^{3+} state (EPR, CT-OAS, and LUMI data); if the heating is performed in a vacuum, such a phenomenon is observed even at temperatures as low as 1500 °C. Octahedrally coordinated Ti^{4+} produces CT absorption bands and luminescence at ~480 nm.

On the other hand, specimens fabricated by the most popular methods (viz., around 2000 °C, in the presence of little or no oxygen), especially by the Verneuil method, show a different OAS profile from that of the specimens including Ti^{4+} considered above. An asymmetric, very weak signal at ~21,000 cm^{-1} (barely visible) and a strong one at ~13,000 cm^{-1} [44] appear, as exemplified in Figure 3, for a transparent spinel ceramic doped with Ti and fabricated under strongly reductive conditions.

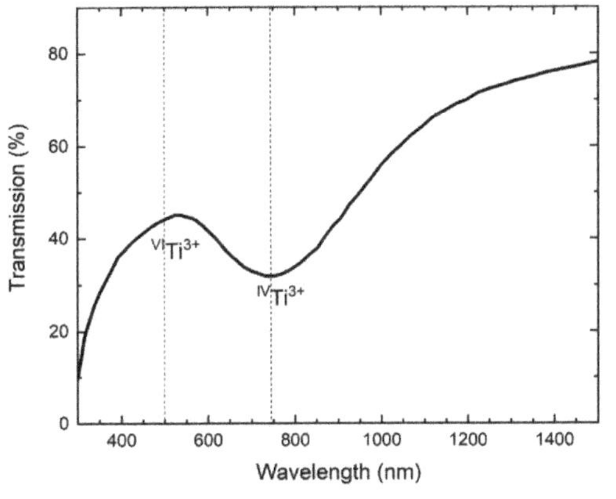

Figure 3. OAS of Ti^{3+} in a transparent ceramic spinel host with a very weak signal of 6-coordinated Ti^{3+} (original data).

The ~21,000 cm^{-1} signal of Figure 3 is assigned, despite its low visibility, to the d-d band produced by single Ti^{3+} cations located in slightly distorted octahedral sites, in analogy with the major band observed in Ti-doped corundum. The strong band in the near-infrared (NIR) domain is assignable to the spin-allowed transition of Ti^{3+} located in a tetrahedral site that is slightly distorted from the cubic symmetry toward D_{2d}. This assignment is considered here as the most likely, although the origin of this band is debated in the literature [45–47]. Indeed, most of Ti^{3+} is likely to replace Mg^{2+} in tetrahedral sites, also because the NIR band appears even in specimens with very low i values, where Al^{3+} antisites are practically absent. EPR signals [46] confirm the existence of some 6-coordinated Ti^{3+}. The Δr between Ti^{3+} and $^{VI}Al^{3+}$ or $^{IV}Mg^{2+}$ is quite similar, while the CFSE for a 6-coordinated cation is somewhat higher. The possible reason for the observed prevalence of a tetrahedral coordination of Ti^{3+} despite the need for charge compensation is discussed below (see below in Section 3.1).

2.2. Vanadium

2.2.1. α-Al_2O_3

V^{5+} is not obtainable in corundum upon doping with vanadium [5]. Similarly, the spectra of V-doped alumina crystals drawn from oxygen-containing melts do not suggest the presence of an oxo-unit [48], although V^{4+} appears in many oxide hosts as VO^{2+} [49]. In corundum, spectra appear to be produced by a simple V^{4+} cation occupying a similar site to that filled by Ti^{3+}.

The species obtained in conventional fabrication conditions is the 6-coordinated V^{3+} ($3d^2$) chromophore. The OAS exhibits three d-d bands (at 17,500, 25,000, and 31,200 cm^{-1}) that can be easily assigned to the $^3T_1$-$^3T_2$, $^3T_1$-$^3T_1(P)$, and $^3A_2$ spin-allowed transitions [5] of a 6-coordinated V^{3+} occupying a site of approximately C_{3v} symmetry.

2.2.2. $MgAl_2O_4$

The stabilization conferred to V^{5+} by a d^0 configuration is very influential, so that noticeable amounts of V^{5+} may be accommodated in suitable fabrication conditions (presence of oxygen, low temperatures) in spinel, YAG [50], or glasses [51]. There are instead no reports suggesting the presence of either VO^{2+} oxo-groups or V^{4+} as such. On the contrary, absorption features that may be associated with d-d transitions are observed at 17,200 and 25,200 cm^{-1} in V-doped spinel [12], as also noted in Table 1. Based on a comparison with alumina or performing band assignment using the Liehr–Ballhausen d^2 diagram [52], the origin of the bands is clearly determined as the $^{VI}V^{3+}$ cation, although a third band predicted at 33,000 cm^{-1} is masked by a CT signal. The relatively high CFSE (0.8×10 Dq, with 10 Dq being ligand field strength in cm^{-1}) is a significant stabilization factor for the d^2 cation.

2.3. Chromium

2.3.1. α-Al_2O_3

The only species evidenced by OAS LUMI and EPR in corundum is 6-coordinated Cr^{3+} [5,16,18,53–58]; indeed, the fluorescence of Cr^{3+} enabled the fabrication of the first actual solid-state laser [58]. All d-d range OAS bands can be assigned to the above-mentioned species, as well as the CT bands in the UV range [35]. In summary, CFSE (1.2×10 Dq), the low Δr with respect to Al^{3+}, and the low IE for the transition from 2+ to 3+ are factors which determine the exclusive preference of corundum for Cr^3.

2.3.2. $MgAl_2O_4$

The speciation pattern is quite similar to that seen in corundum. The only species identifiable by spectroscopy is 6-coordinated Cr^{3+} [17,19]; for comparison, Cr^{6+} is the dominant or exclusive species in glass at similar or higher basicity [59]. The OAS of Cr^{3+} located in a spinel ceramic specimen appears in Figure 4. Red photoluminescence can be recorded for such a sample, just like in corundum [60]. The stability of Cr^{3+} is not sensitive to the redox characteristic of the atmosphere during thermal treatments. As for $^{IV}Cr^{3+}$, it was not observed in spinel. This may be due to the large CFSE of a d^3 cation (1.2×10 Dq), which drops to 0.36×10 Dq in tetrahedral sites.

2.4. Manganese

2.4.1. α-Al_2O_3

Unsurprisingly, Mn^{3+} was found in view of its d-d band at 20,600 cm^{-1} (^5E_g-$^5T_{2g}$) in OAS [5] to be the only species in impurity-free corundum crystals. However, single crystals formed from a PbO-containing flux at around 1000 °C also contain some Mn^{4+} [61]; in practice, such a method also introduces some Pb^{2+} in the crystal. Mn^{3+} was also detected by OAS (besides Mn^{2+} and Mn^{4+}) as the dominant species in corundum powder prepared from clay in supercritical conditions, as long as other Me^{2+} impurities were absent [62]. In practice, Mn^{4+} was successfully introduced by adding charge compensators like Mg^{2+}, if atmospheres containing relatively high oxygen concentration were employed [63–65]. For instance, Mn-doped alumina crystals grown in air at low temperatures with simultaneous introduction of charge compensators [64,65] contained Mn^{4+} as the dominant or exclusive dopant.

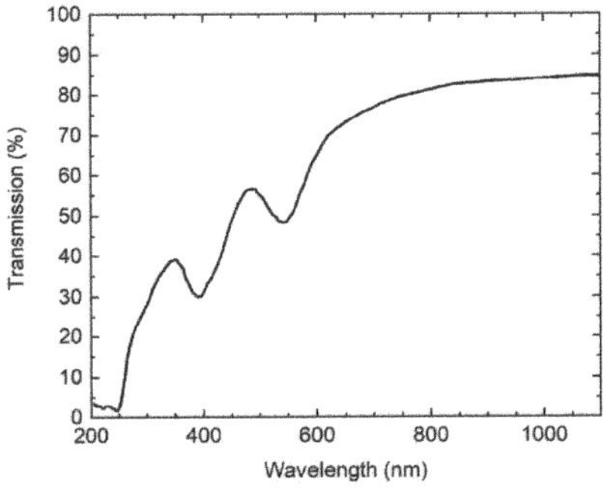

Figure 4. OAS of a Cr^{3+}-doped transparent ceramic spinel (original data).

Compared to $^{VI}Al^{3+}$, $^{VI}Mn^{2+}$ is a large ion (a radius of 67 pm even in ls state). It has, however, a stable d^5 configuration. Empirical data strongly suggest that the stabilization energy associated with this feature compares to that imparted to a d^3 configuration by CFSE, supported also by the relatively large IE for the 2+ to 3+ jump. Fabrication of both Mn-doped crystals and transparent ceramics usually involves significantly reducing atmospheres; for single crystals, temperatures around 2000 °C are also necessary. There are therefore a few factors acting in favor of Mn^{2+}, while the radii difference remains a drawback. Experiments have shown that the cation could be introduced in small amounts in an artificial corundum crystal fabricated under a very reductive atmosphere [22]. As indicated above, powder prepared in supercritical conditions and avoiding the presence of Me^{2+} impurities also contains some Mn^{2+} [62], identified quite clearly by its EPR signature, while OAS and LUMI signals are masked by those of other states. The identification of Mn^{2+} in an alumina crystal by EPR is also reported by Liebach et al. [20], whose synthesis method yielded mixed oxidation states by using a PbO-PbF_2 melt (favoring a low oxidation state) at low temperatures (around 1000 °C, favoring the appearance of Mn^{3+}) in air. The OAS spectrum indeed shows weak signals pertaining to a mixture of Mn^{2+} and Mn^{3+}.

Overall, it can be concluded that 6-coordinated Mn^{3+} is the prevalent state assumed by Mn in pure corundum under mainstream fabrication conditions; other Mn species may enter the lattice only under special conditions.

2.4.2. $MgAl_2O_4$

The accommodated oxidation states depend on fabrication conditions, like for alumina [21,23,66–68]. However, only Mn^{2+} and Mn^{3+} can be accommodated in spinel. In ceramic $MgAl_2O_4$ sintered in air at 1600 °C and then subjected to hot isostatic pressing under Ar at ~1500 °C, yellow-to-orange or brown hues are obtained as the Mn content increases. The OAS shown in Figure 5a (previously unpublished data) indicates the presence of both Mn^{2+} and Mn^{3+}; while not obvious at first glance, closer inspection reveals similarities with the spectrum reported by Liebach et al. [20]. Similarly, Mn^{3+} was also detected in $MgAl_2O_4$ single crystals with an i = 0.23, prepared in an oxide flux at low temperatures ($Na_2B_4O_7$; 1200 + 800 °C) [69]; the color evolved like in the ceramic sample as a function of Mn content. Mn^{3+} substitution of Al^{3+} was the main process caused by doping, but some Mn^{2+} formed as well. After Mn^{3+} incorporation, the symmetry of the Al^{3+} site was distorted, becoming more similar to that of corundum.

In the vast majority of conditions, however, 4-coordinated Mn^{2+} sitting in slightly distorted sites is the stable oxidation state in $MgAl_2O_4$ spinel crystals, accompanied by a minor fraction of 6-coordinated Mn^{2+}. The 6-coordinated species is promoted by an increase in Mn dopant concentration and in alumina content of the spinel host (as in non-stoichiometric solid solutions). In Figure 5b, the typical OAS seen in a Mn-doped spinel crystal is presented, with Mn^{2+} being 4-fold coordinated.

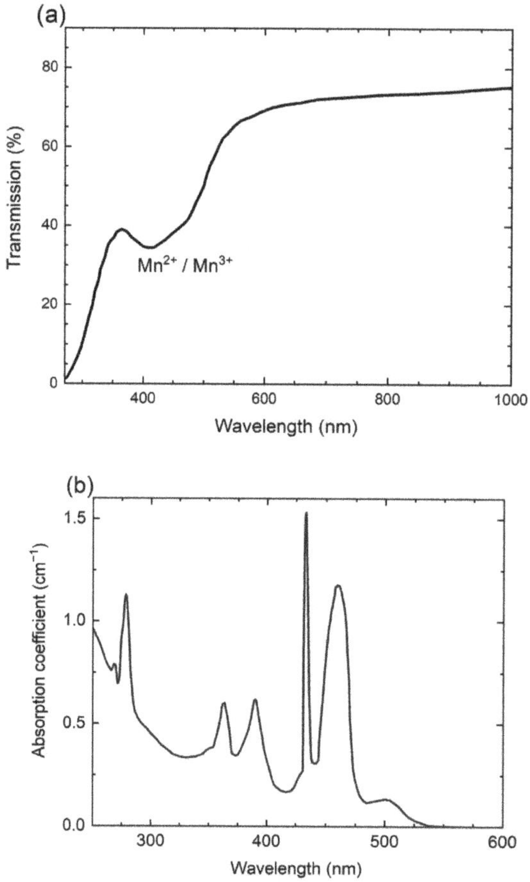

Figure 5. OAS of Mn-doped spinel ceramics and single crystal: (**a**) OAS of Mn-doped transparent spinel ceramics showing the presence of Mn^{2+} and Mn^{3+} (original data); (**b**) OAS of a Mn-doped corundum single crystal (original data).

2.5. *Iron*

2.5.1. α-Al_2O_3

Despite the impressive amount of work dedicated to elucidating how Fe speciates in corundum, the issue is still under debate [5,24,70–81]. It is clear that the speciation pattern significantly depends on fabrication (or formation, in the case of natural sapphire) conditions, total concentration of Fe, and presence of impurities, especially Ti. $^{VI}Fe^{3+}$ is clearly the dominant (or unique) species in atmospheres including low concentrations of oxygen. In the case of specimens formed under reductive or very reductive conditions, a strong additional absorption band appears at 11,500 cm^{-1}. In natural crystals, this NIR band is accompanied by a contiguous, broad, and flat absorption centered at ~17,000 cm^{-1} [75,78,81]. Some authors are inclined to see the NIR band as stemming from the presence of $^{VI}Fe^{2+}$, while assigning the absorption in the yellow domain to homo-polynuclear complexes linking Fe^{2+} to Fe^{2+}, Fe^{3+} to Fe^{3+}, and/or Fe^{2+} to Fe^{3+} [78–81]. A strong band clearly assigned to $^{VI}Fe^{2+}$ [82,83] is indeed found in glass at around 10,000 cm^{-1}, although this is not accompanied by absorption in the yellow domain. In artificial transparent ceramics of Fe-doped spinel, this NIR band pertaining to $^{VI}Fe^{2+}$ is the only signal present, as shown in Figure 6 (original data).

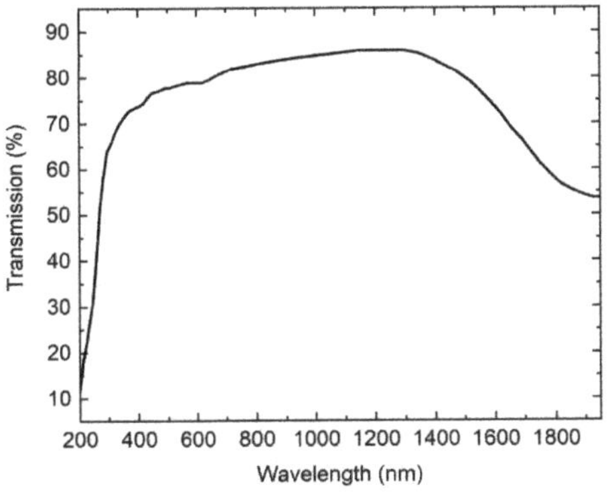

Figure 6. OAS showing the d-d band of Fe^{2+} observable in an Fe-doped ceramic spinel host (original data).

No indication of the presence of $^{IV}Fe^{3+}$ (or $^{IV}Fe^{2+}$) is given by the OASs of Fe-doped alumina. In corundum and other oxide hosts, the band in the yellow domain was assigned by Sherman to a hetero-polynuclear complex including Fe^{2+} and Ti^{4+} impurities [84]. He specifically argued that molecular orbitals (ensuring a weak bond) can form more easily in these structures because the involved cations lie in face-sharing octahedral sites. In such clusters, charge transfer occurs from Fe^{2+} to Ti^{4+}. His molecular orbital calculations predict a band at ~18,000 cm^{-1}, reasonably close to observations. The two cations indeed offer t_{2g} symmetry orbitals, which may be superposed. This theory is the most popular today; while such clusters are likely to exist in certain Ti- and Fe-containing sapphires, their blue coloration may also be the result of other mechanisms.

For the sake of completeness, let us note that 5-coordinated Fe^{3+} (observed in certain oxide crystals [26]) was not mentioned in the literature, for the case of corundum; we also find such an assumption dubious in the case of glass, contrary to other authors (e.g., [85]).

2.5.2. $MgAl_2O_4$

The speciation pattern of Fe in spinel is markedly different from that observed in corundum. In a colorless MgO-$3.5Al_2O_3$ crystal host, the d-d spectrum consists of a peak-and-shoulder system extending in the range 3800 to 5500 cm^{-1} and exhibiting an oscillator strength typical for complexes lacking an inversion point. This is characteristic of an Fe^{2+} cation located in a slightly distorted tetrahedral complex, also because 6-coordinated Fe^{2+} absorbs at much higher frequencies, such as at ~10,000 cm^{-1} in glass. The UV absorption cut-off advances till ~25,000 cm^{-1} into the visible range, indicating intense CT-related absorption. Analysis of the CT absorption does not offer unambiguous results but suggests the presence of some Fe^{3+}. Modeling of Fe-doped stoichiometric spinel also led De Souza et al. [86] to predict the presence of some 6-coordinated Fe^{3+} in Al^{3+} sites, along the dominant Fe^{2+}. Fe^{3+} is detected in opaque ceramics sintered in air at low temperatures, mostly as 6-coordinated species but accompanied by some $^{IV}Fe^{3+}$ [87]. The presence of Fe^{3+} as a single cation and clusters is also proposed in [88] for Fe-doped synthetic spinels, based on the interpretation of OAS and Mössbauer data. In these works, no Fe^{2+} is mentioned but the recorded OAS do not extend to wavelengths higher than ~780 nm. In Fe-rich natural $MgAl_2O_4$ crystals and in $FeAl_2O_4$ [89–91], the presence of some $^{VI}Fe^{2+}$ has been demonstrated, besides the dominant 4-coordinated Fe^{2+} cations. In [92], no Fe^{3+} was detected, while the Fe-containing spinel crystals examined in [93,94] mostly contained $^{IV}Fe^{2+}$ with some 6-coordinated Fe^{2+} (10–15% of total divalent iron; a few Fe^{2+}-Fe^{3+} clusters and even some $^{IV}Fe^{3+}$ were seen, with signals in the 18,000–21,300 cm^{-1} range). Slack's data [89] offer a similar picture to that of [93], with the splitting of the IR band related to 4-coordinated Fe^{2+} being emphasized.

In summary, 6-coordinated Fe^{3+} dominates in alumina especially at high dopant concentrations, but some octahedral Fe^{2+} may be stabilized in suitable conditions; no 4-coordinated species are observed. The Fe^{2+}/Fe^{3+} ratio is much larger in spinel, with Fe^{2+} mostly seeking the tetrahedral sites.

2.6. Cobalt

2.6.1. α-Al_2O_3

Spectral examination [5] indicates Co^{3+} as being the virtually unique TM+ present in Verneuil crystals. However, Co^{3+} can be partially reduced to 6-coordinated Co^{2+} above 1500 °C in H_2-containing atmospheres [25,95]. Data on Co^{2+} in corundum also appear in [96–103].

2.6.2. $MgAl_2O_4$

As in the case of Fe, speciation in spinel follows a clearly distinct pattern from that seen in corundum. An example of OAS for a Co^{2+}-containing transparent spinel ceramic appears in Figure 7 (original data). It is difficult to detect 6-coordinated Co^{2+} by OAS when the tetrahedral species is present; some authors have considered that some $^{VI}Co^{2+}$ may also exist [26,28,41,104–106].

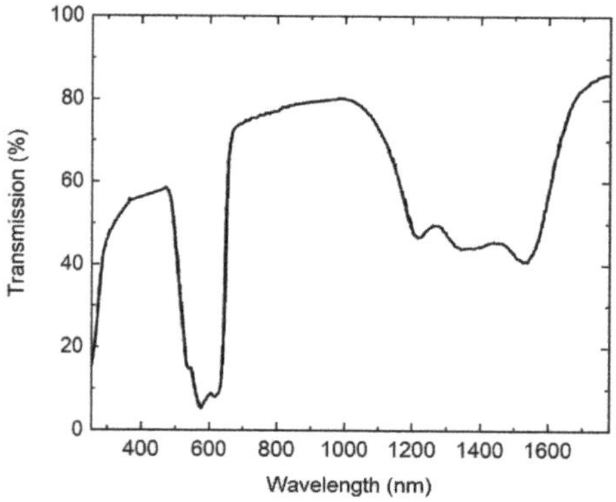

Figure 7. OAS of a Co-doped transparent spinel ceramic (original data).

$^{VI}Co^{3+}$ is a species that can be transiently obtained in a host like $CoAl_2O_4$, if annealed at around 500 °C under oxidizing conditions [107]. Some $MgCo_2O_4$ was found, besides other phases, when CoO_3 reacted with $MgAl_2O_4$ for catalyst synthesis. Based on the above, this species is not listed among those accommodable in transparent spinel in Tables 1–3.

Comparing spinel and corundum, 6-coordinated Co^{3+} appears to be the preferred species in corundum, whereas 4-coordinated Co^{2+} is readily accepted by spinel, although some 6-coordinated Co^{2+} may appear.

2.7. Nickel

2.7.1. α-Al_2O_3

Assignment of the bands most commonly appearing in OAS and EPR data of Ni-doped corundum is detailed in [5,27,108–110]. The work of Solntsev et al. [110] on Ni^{3+}-containing chrysoberyl is particularly helpful for band assignments in corundum, as it shows the three lowest spin-allowed transitions of the $3d^7$ cation at 9100, 18,300, and 23,000 cm^{-1}. In any case, the exclusive presence of Ni^{3+} sitting in octahedral sites is clear. Only under special conditions (e.g., heating in a vacuum and coating with NiO) is it possible to reduce some Ni^{3+} to Ni^{2+}; here, the Δr problem mentioned above for the cobaltous cation is even more acute.

2.7.2. $MgAl_2O_4$

Blue Ni-doped spinel crystal or ceramics were found to carry only Ni^{2+} [110–112], whereas Ni^{3+} was present in corundum [5,27]. A final decision on the coordination type(s) of Ni^{2+} is somewhat complicated. For instance, Kuleshov et al. assigned all OAS bands [28] to 6-coordinated Ni^{2+}; they also detected emission at ~8100 cm^{-1}. However, Reed and Kay saw a dominant $^{IV}Ni^{2+}$, with only one band suggesting some $^{VI}Ni^{2+}$ [12]. In the detailed spectrum of Ni^{2+}-doped spinel, they recorded bands at 4500 cm^{-1} (4-coordinated), 9500 cm^{-1} (4- and 6-coordinated), 13,500 with a weak shoulder at 14,500 cm^{-1} (4- and 6-coordinated), 21,000 cm^{-1} (4-coordinated), 23,000 cm^{-1} (4- and 6-coordinated), and 27,000 cm^{-1} (6-coordinated). Sakurai et al. [112] made a similar observation.

We similarly believe that Ni^{2+} exists in both roughly octahedral and tetrahedral sites; however, the dominant species is 4-coordinated Ni^{2+}. This dominance is better identified in ceramic spinels doped with Ni, such as the specimen whose OAS is presented in Figure 8. The signature of the octahedral species appears as a wide, weak signal at ~13,500 cm^{-1} relating to the transition from $^3A2$ to $^3T1(F)$; it is possible that the ~27,000 cm^{-1} band could also belong to the 6-coordinated species as $^3T1(P)$.

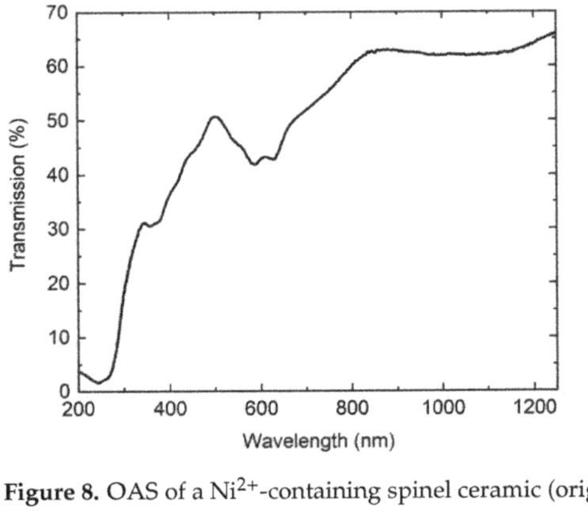

Figure 8. OAS of a Ni^{2+}-containing spinel ceramic (original data).

2.8. Copper

2.8.1. α-Al_2O_3

It appears that an insertion of Cu in corundum is difficult for crystals grown in Verneuil conditions [5]. At lower temperatures (1200 °C) and in oxygen-bearing atmospheres, copper enters the lattice exclusively as $^{VI}Cu^{3+}$ [113]. The picture provided by the optical and EPR data [113–115] reveals a d^8 cation located in an octahedral site, within which the ion migrates slightly along the C_3 axis. The elongated octahedron geometry site retains trigonal rotational symmetry; therefore, either C_{3v} or D_3 approximate symmetry were considered in calculations. The s/o interaction is noticeable (ζ = ~600 cm^{-1}). The most complete calculation [114,116], which also considered ligand effects, assigned the three spin-allowed optical bands as follows: 17,500 cm^{-1} ($^3A_2$ to $^1E(D)$), 21,000 cm^{-1} ($^3T_{2g}$ (F)), and 31,000 cm^{-1} ($^3T_{1g}$(F), probably also including $^1A_1$(G)), with a covalency factor k = 0.68. Petrosyan et al. [115] noted that the interaction with the ground state (under s/o) derived from a few (and not only the first) excited levels needs to be factored in for the best fit of the paramagnetic parameters (g_{II} = 2.0988, $g_{\perp}$ = 2.0973; D = −0.884).

Under drastic reductive conditions (H_2 for a few hours at 1200 °C), the d^8 dopant could be reduced to $^{VI}Cu^{2+}$. In [29], an emission band attributed to Cu^{2+} was detected at ~8000 cm^{-1}. While the EPR signals recorded had some problems, the authors observed features indicating a tetragonal J-T distortion of the Al^{3+} sites (initially of C_{3v} symmetry) upon replacement with Cu^{2+}. It was also shown that doping corundum with Ge^{4+} facilitates the stabilization of Cu^{2+}. There are also other studies regarding the LUMI and EPR of Cu^{2+} adsorbed on the surface of alumina grains [31,117]. However, it is not clear to what extent the complexes formed in the latter work are identical to those in which Cu^{2+} enters bulk crystal sites. It is maybe worth noting that Li et al. [117] detected a further tetragonal distortion of the Al^{3+} site upon accommodation of Cu^{2+}. Based on radius considerations, it seems quite unlikely that Cu^+ may be accommodated in alumina.

2.8.2. $MgAl_2O_4$

The examination of OAS produced by Cu-doped spinel crystals or ceramics shows that no $^{VI}Cu^{3+}$ (dominant species in corundum) exists here. We have made (using the d^2 T-S diagram) an approximative calculation of the band positions for hypothetical $^{IV}Cu^{3+}$ located in an undistorted tetrahedral site. Comparison with the above-shown OAS ascertains that 4-coordinated Cu^{3+} is also absent.

In Figure 9a,b, the OAS of two Cu-doped transparent ceramic spinel specimens is presented (original data). In Figure 9a, the top curve has a faint signal at ~720 nm, indicating the presence of a small amount of 6-coordinated Cu^{2+}, also detected by [30]. In Figure 9b, the top curve shows a band in the NIR, also shown by [30]. The band, as noted above, is generated by a 4-coordinated Cu^{2+}. The cation producing this band is located in a slightly compressed tetrahedral site (the actual symmetry is orthorhombic D_2 or tetragonal D_{2d}). In the lower curve of Figure 9a, a weak signal appears at 590 nm. It can be assigned with confidence to traces of Cu^0 based on [118,119].

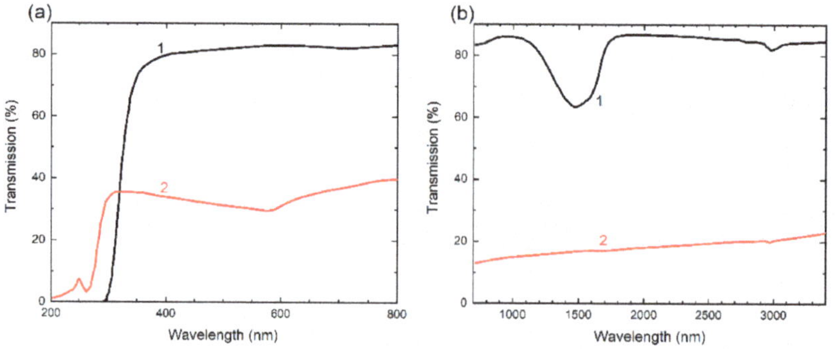

Figure 9. OAS of Cu^{2+} containing transparent $MgAl_2O_4$ spinel disks, corresponding to the bulk of the specimens after grinding off the red (Cu^0) coatings. Top curves (1) correspond to a greenish sample with 0.3% CuO; bottom curves (2) correspond to a colorless sample (with traces of red) containing 0.1% CuO. (**a**) UV-VIS domain and (**b**) segment of IR domain. Reproduced with permission from unpublished work at Fraunhofer IKTS in Dresden (joint research with A. Goldstein).

3. Discussion

An analysis of the spectral data enabled us to establish the speciation patterns of 3d elements in the examined hosts with a high level of confidence. Below, the correlation of these patterns with the most likely controlling factors is discussed.

3.1. Preference of Late 3d Transition Metals for Adopting the 2+ Oxidation State and Tetrahedral Sites in Spinel

This correlation constitutes the most interesting finding of this study. The data show that the 3+ state is strongly preferred in alumina throughout the 3d series. The 2+ state may also be stabilized in the octahedral sites of alumina, but only in the case of late 3d elements and if specimens are fabricated under strongly reductive conditions. In spinel, the late 3d transition metals also assume the 2+ state in specimens fabricated under a wide range of atmospheres, including moderately oxidative conditions (extremely oxidative conditions like those used in [59] have not yet been examined). In addition to the 2+ state, a tetrahedral coordination is favored in spinel, even by Ti^{3+} and Ni^{2+} (despite a CFSE in 6-fold coordination being significantly higher than in 4-fold coordination for the latter).

The reasons for the observed behavior are not straightforward. However, these observations may be rationalized considering the interaction of several factors. First of all, it is known and noted above that IE increases along the 3d series, with a slight drop from Mn to Fe. This increase significantly reduces the ability of the late 3d elements to witness a rise in their oxidation state from 2+ to 3+, for instance when an increase in the oxidative power of the atmosphere requires such a shift. This phenomenon is active in both spinel and alumina, but structural dissimilarities lead to different results. In corundum, all octahedral sites occupied by native cations (viz., those allowing preferential substitutional exchange based on Madelung considerations) harbor 3+ cations. On the contrary, one-third of the non-vacant sites of the spinel structure are tetrahedral and occupied by 2+ native cations. The latter offers an additional option for doping with TMs. Other factors of influence notwithstanding, the trend defined by the IE makes these sites the most attractive. In addition, we saw that IE must act in synergy with other factors for the incorporation of 2+ cations in corundum, for instance in combination with highly reductive atmospheres during high-temperature thermal treatments; this observation is valid for both single crystals and ceramic fabrication. This is because such atmospheres favor low oxidation states, providing a mechanism for the compensation of the charge on the TM'_{Al} defects. A possible scenario for the process occurs in two stages. A reduction in the oxygen concentration in the furnace tilts the process depicted by Equations (2) and (3) to the right:

$$O_O \leftrightarrow \tfrac{1}{2}\, O_2\ (g) + V_{\ddot{o}} + 2\, e^- \tag{2}$$

$$K_n = n^2 \cdot [V_{\ddot{o}}] \cdot p_{O2}^{\ 0.5} \tag{3}$$

In the second stage, a TM cation with a charge larger than 2+ can be reduced by the electrons trapped at the vacancy site. The vacancies, thus freed, offer charge compensation for the formed TM'_{Al} defects. The effect of IE is also modulated by Δr, as most $^{IV}TM^{2+}$ radii are close to that of $^{IV}Mg^{2+}$, while $^{VI}TM^{2+}$ species differ more markedly from $^{VI}Mg^{2+}$ or $^{VI}Al^{3+}$ and would require charge compensation.

The dopants may maintain a 2+ state as induced by the IE by replacing both octahedrally and tetrahedrally coordinated native cations in spinel. Data show that Δr and even the necessity for charge compensation are not decisive in the choice for the site type, as the optimization of intra-complex negative charge by switching from octahedral to tetrahedral coordination seems more important. For instance, it was observed that ligand coordination numbers change from 6 to 4 in glass in order to better accommodate the increase in intra-complex negative charge [33,72]. TMs in the range from Ti to Fe respond by accommodating higher TM dopant oxidation states when the charge on the ligands increases in glasses. This ceases to be the case with Co. More accurately, Co may progress from 2+ to 3+, but only at ultra-high levels of basicity [72], not encountered in normal glasses. Even within the Ti to Fe range, Mn exhibits the adoption of 4-fold coordination for Mn^{2+}, to some extent. In the case of the latest 3d elements (Ni, Cu), oxidation does not occur even for maximal basicity level. The fact that Co^{2+} reduces its coordination as the concentration of alkaline ions increases (in other terms, as basicity increases) was observed more than one hundred years ago [104]. Weyl even suggested that the increased ligand size (accommodating more negative charge as $\delta-$ increases) allowed for the adoption of the more compact tetrahedral site instead of the octahedron. We note these observations because this process is also relevant here, in our view. The $\delta-$ on the oxide ions of both of our hosts is quite large (being somewhat higher in corundum). In pure corundum (as opposed to spinel), $\delta-$ can be estimated by the method developed by Duffy for glasses, obtaining a theoretical basicity $\Lambda_{theo} = 0.61$ that is similar to that of a binary $70SiO_2$-$30Na_2O$ glass (i.e., larger than the one that prompts Co^{2+} to change its coordination in glass). In order to compensate such a $\delta-$ level, oxidation higher than 2+ would be a good solution for the dopants. However, the late TMs (including Co) are forced to stay at a 2+ level due to IE and find another solution to optimize the negative charge distribution, i.e., a reduction in coordination number. To understand this latter mechanism, it is important to note that, for a given $\delta-$ and TM+, the TM+-$O^{\delta-}$ distance is shorter in a tetrahedral complex than in an octahedral one, favoring covalency. Especially for π-bonds, a better orbital superposition can be achieved (*p* orbitals from ligands and *d* orbitals from central cations). A covalency increase causes a displacement of charge from the ligands toward the central cation (i.e., the nephelauxetic effect [120]); a complex with a negative charge therefore redistributed is a system of lower electrostatic energy. This means that the tetrahedral TM^{2+} complexes allow for a more judicious accommodation of the negative charge than an octahedral antisite, for instance. The preference for tetrahedral sites is also exhibited by the Ti^{3+} cation, viz., an element for which substituting a 6-coordinated Al^{3+} would mean elimination of the necessity for charge compensation.

In the case of Ni^{2+}, the strength of the CFSE (1.2×10 Dq, not equaled by any of the neighboring elements) leads to the formation of some 6-coordinated species but the dominant species remains the 4-coordinated one.

3.2. *Comparison of the Attractivity of Normal Sites and Antisites for Doping Transition Metal Cations in Spinel*

Above, we considered tetrahedral Mg^{2+} sites in spinel specimens as the only possible target for incoming TM cations. This is acceptable when $i = 0$, viz., a rare case. For $i \neq 0$, Mg^{2+} antisites also exist. Such charged point defects, however, are configured as dipoles with those formed as Al'_{Mg}. An energetically costly lattice rearrangement is needed in order to disengage the Mg^{2+} antisites and thereby make an octahedral site available to an incoming 2+ doping cation. We speculate that, owing to the above, the antisites are not involved in accommodating the dopants.

3.3. *Entrance of Transition Metal Cations with an Oxidation Number Higher than 3+*

Though of secondary importance, this issue merits a few words. The highest oxidation state of the native cations in corundum and spinel is 3+. Nonetheless, Ti^{4+} can be introduced in both hosts because a few favorable factors work in tandem for this specific ion. Compared to later TMs, the 3+ state of Ti needs to be exceeded by only one charge unit in order to reach the high-stability $3d^0$ configuration corresponding to Ti^{4+}. The IE pressure opposing the oxidation state increase is low for the early 3d elements. Moreover, the size difference with Al^{3+} is only 9%.

V^{5+} (d^0) was also observed but only in a spinel host, suggesting a different behavior of the two lattices under study. However, such a conclusion may be premature, as V-doped corundum specimens fabricated in conditions that would be favorable for V^{5+} have not yet been examined. Our prediction is that V^{5+} will also prove accommodable within corundum, if under suitable conditions. Other high-oxidation TM+ cations corresponding to a d^0 configuration do not appear in corundum nor spinel. This is not surprising because, as one moves to the right of V, the oxidation states for attaining a d^0 state increase (Cr^{6+}, Mn^{7+}, etc.). The small radii, decreasing proportionally to the oxidation state increase, also move toward 4-fold coordination and thus toward a Δr increase with respect to Al^{3+}. One additional reason for the difficult stabilization of high-oxidation states is the fact that charge compensation requires expulsion from the lattice of strongly bonded cations (to "free" negative charge) and/or very high values of oxygen gas pressure. For the reasons presented above, high-oxidation-state dopants are rare. However, Mn^{4+} represents another notable exception: due to its maximal CFSE of 1.2×10 Dq and a similar size to Al^{3+}, this cation can be stabilized in corundum under certain fabrication conditions.

4. Conclusions

The speciation patterns of the 3d elements in spinel and corundum were established by analyzing their spectral data. The correlation between the main factors controlling the speciation process and the patterns observed was examined; certain rules regarding the way it plays out were discovered. The speciation process differed significantly in the two hosts examined.

The main observed differences in the speciation patterns caused by structural dissimilarities between corundum and spinel lattices are as follows:

- Spinel has a tendency to accommodate incoming TMs located in the second half of the 3d series in a 2+ state in its tetrahedral sites, which typically harbor 2+ native cations.
- Corundum prefers to accommodate 3+ TM+; 2+ cations are difficult to form, but they may be stabilized in the case of late TMs if suitable processing conditions are selected.
- A strong motive for choosing tetrahedral sites is the ability of TM+ cations to distribute the intra-complex negative charge in a more judicious way.
- The conclusions summarized above hold true when the compounds α-Al_2O_3 and $MgAl_2O_4$ are compared. It is still to be determined whether they may possess more general applicability for corundum and spinel structures obtained in different compositional systems.

Author Contributions: Conceptualization, A.G.; validation, A.G. and A.Z.; investigation, A.G.; data curation, A.G.; writing—original draft preparation, A.G.; writing—review and editing, A.Z.; visualization, A.Z. All authors have read and agreed to the published version of the manuscript.

Funding: This research received no external funding.

Data Availability Statement: No new data were created or analyzed in this study.

Conflicts of Interest: The authors declare no conflicts of interest.

References

1. Goldstein, A.; Krell, A.; Burshtein, Z. *Transparent Ceramics: Materials, Engineering, and Applications;* John Wiley & Sons: New York, NY, USA, 2020.
2. Henderson, B.; Bartram, R.H. *Crystal-Field Engineering of Solid-State Laser Materials;* CUP: Cambridge, UK, 2005.
3. Figgis, B.N. *Introduction to Ligand Fields;* Interscience Publishers: New York, NY, USA, 1966.

4. Zandonà, A.; Castaing, V.; Shames, A.I.; Helsch, G.; Pirri, A.; Toci, G.; Deubener, J.; Allix, M.; Goldstein, A. Effect of the interaction between basicity and reductive character of melting atmosphere—Both extreme—On the oxidation and coordination states assumed by transition metals when doped to silicate glasses. *J. Non-Cryst. Solids* **2024**, *637*, 123038. [CrossRef]
5. McClure, D.S. Optical Spectra of Transition-Metal Ions in Corundum. *J. Chem. Phys.* **1962**, *36*, 2757–2779. [CrossRef]
6. Aramburu, J.A.; Garcia-Fernandez, P.; García-Lastra, J.M.; Barriuso, M.T.; Moreno, M. Internal electric fields and color shift in Cr3+-based gemstones. *Phys. Rev. B* **2012**, *85*, 245118. [CrossRef]
7. Bristow, J.K.; Parker, S.C.; Catlow, C.R.A.; Woodley, S.M.; Walsh, A. Microscopic origin of the optical processes in blue sapphire. *Chem. Commun.* **2013**, *49*, 5259–5261. [CrossRef]
8. Gaudry, E.; Kiratisin, A.; Sainctavit, P.; Brouder, C.; Mauri, F.; Ramos, A.; Rogalev, A.; Goulon, J. Structural and electronic relaxations around substitutional Cr^{3+} and Fe^{3+} ions in corundum. *Phys. Rev. B* **2003**, *67*, 094108. [CrossRef]
9. Sickafus, K.E.; Wills, J.M.; Grimes, N.W. Structure of Spinel. *J. Am. Ceram. Soc.* **1999**, *82*, 3279–3292. [CrossRef]
10. Mo, S.-D.; Ching, W.Y. Electronic structure of normal, inverse, and partially inverse spinels in the $MgAl_2O_4$ system. *Phys. Rev. B* **1996**, *54*, 16555–16561. [CrossRef]
11. O'Neill, H.S.C.; Navrotsky, A. Simple spinels; crystallographic parameters, cation radii, lattice energies, and cation distribution. *Am. Mineral.* **1983**, *68*, 181–194.
12. Reed, J.S.; Kay, H.F. Optical Spectra of 3d Transition Metal Ions in MgO $3{\cdot}5_2O_3$ Spinel. *J. Am. Ceram. Soc.* **1969**, *52*, 307–311. [CrossRef]
13. Yamaga, M.; Yosida, T.; Hara, S.; Kodama, N.; Henderson, B. Optical and electron spin resonance spectroscopy of Ti^{3+} and Ti^{4+} in Al_2O_3. *J. Appl. Phys.* **1994**, *75*, 1111–1117. [CrossRef]
14. Lombard, P.; Boizot, B.; Ollier, N.; Jouini, A.; Yoshikawa, A. Spectroscopic studies of Ti^{3+} ions speciation inside $MgAl_2O_4$ spinels. *J. Cryst. Growth* **2009**, *311*, 899–903. [CrossRef]
15. Powell, R.C.; Caslavsky, J.L.; AlShaieb, Z.; Bowen, J.M. Growth, characterization, and optical spectroscopy of $Al_2O_3{:}Ti^{3+}$. *J. Appl. Phys.* **1985**, *58*, 2331–2336. [CrossRef]
16. Reddy, S.L.; Reddy, K.N.M.; Reddy, N.C.G.; Frost, R.L.; Endo, T. Electron paramagnetic resonance spectral studies on natural sapphire. *Mol. Phys.* **2009**, *107*, 1537–1539. [CrossRef]
17. Singh, V.; Chakradhar, R.P.S.; Rao, J.L.; Kim, D.-K. Combustion synthesized $MgAl_2O_4$:Cr phosphors—An EPR and optical study. *J. Lumin.* **2009**, *129*, 130–134. [CrossRef]
18. Adachi, S. Review—Photoluminescence Properties of Cr^{3+}-Activated Oxide Phosphors. *ECS J. Solid State Sci. Technol.* **2021**, *10*, 026001. [CrossRef]
19. Hao, Y.; Wang, S.; Zhang, Y. Effect of Cr^{3+} on the microstructure and photoluminescence of $MgAl_2O_4$ transparent ceramic. *J. Lumin.* **2022**, *242*, 118542. [CrossRef]
20. Liebach, R.A.; Dobbie, J.; Hutton, D.R.; Troup, G.J.F. ESR and optical spectra of Mn^{2+} sapphire. *J. Gemm.* **1988**, *21*, 227–231. [CrossRef]
21. Clausen, R.; Petermann, K. Mn^{2+} as a potential solid-state laser ion. *IEEE J. Quantum Electron.* **1988**, *24*, 1114–1117. [CrossRef]
22. Low, W.; Suss, J.T. Paramagnetic Resonance Spectrum of Manganese in Corundum. *Phys. Rev.* **1960**, *119*, 132–133. [CrossRef]
23. Shaffer, J.S.; Farach, H.A.; Poole, C.P. Electron-spin-resonance study of manganese-doped spinel. *Phys. Rev. B* **1976**, *13*, 1869–1875. [CrossRef]
24. Krebs, J.J.; Maisch, W.G. Exchange Effects in the Optical-Absorption Spectrum of Fe^{3+} in Al_2O_3. *Phys. Rev. B* **1971**, *4*, 757–769. [CrossRef]
25. Müller, R.; Günthard, H.H. Spectroscopic Study of the Reduction of Nickel and Cobalt Ions in Sapphire. *J. Chem. Phys.* **1966**, *44*, 365–373. [CrossRef]
26. Shirakov, A.; Burshtein, Z.; Goldstein, A.; Frumker, E.; Ishaaya, A. Use of Co:$MgAl_2O_4$ transparent ceramics in passive Q-switching of an Er:Glass laser at 1.534 µm. *Opt. Express* **2020**, *28*, 21956–21970. [CrossRef]
27. Geschwind, S.; Remeika, J.P. Spin Resonance of Transition Metal Ions in Corundum. *J. Appl. Phys.* **1962**, *33*, 370–377. [CrossRef]
28. Kuleshov, N.V.; Shcherbitsky, V.G.; Mikhailov, V.P.; Kück, S.; Koetke, J.; Petermann, K.; Huber, G. Spectroscopy and excited-state absorption of Ni^{2+}-doped $MgAl_2O_4$. *J. Lumin.* **1997**, *71*, 265–268. [CrossRef]
29. Romanov, A.N.; Haula, E.V.; Kapustin, A.A.; Kostyukov, A.A.; Egorov, A.E.; Kuzmin, V.A.; Korchak, V.N. Broadband near infrared photoluminescence of Cu^{2+}-doped corundum (α-Al_2O_3). *J. Am. Ceram. Soc.* **2024**, *107*, 979–983. [CrossRef]
30. Reed, J.S. Optical Spectra of Cu^{2+} and Ni^{2+} in Intermediate Aluminate Spinels. *J. Am. Ceram. Soc.* **1973**, *56*, 525–527. [CrossRef]
31. Hronský, V.; Rákoš, M. Epr study of Cu^{2+} ions adsorption on aluminium oxide. *Czechoslov. J. Phys. B* **1983**, *33*, 1347–1356. [CrossRef]
32. Winkler, E.R.; Sarver, J.F.; Cutler, I.B. Solid Solution of Titanium Dioxide in Aluminum Oxide. *J. Am. Ceram. Soc.* **1966**, *49*, 634–637. [CrossRef]
33. Lacovara, P.; Esterowitz, L.; Kokta, M. Growth, spectroscopy, and lasing of titanium-doped sapphire. *IEEE J. Quantum Electron.* **1985**, *21*, 1614–1618. [CrossRef]

34. Daimon, T.; Naruse, H.; Watanabe, H.; Oda, H.; Yamanaka, A. Study of Blue Photoluminescence in Titanium Doped Al_2O_3 Single-Crystals. *IOP Conf. Ser. Mater. Sci. Eng.* **2011**, *18*, 102012. [CrossRef]
35. Tippins, H.H. Charge-Transfer Spectra of Transition-Metal Ions in Corundum. *Phys. Rev. B* **1970**, *1*, 126–135. [CrossRef]
36. Gladney, H.M.; Swalen, J.D. Theory of EPR of Ti^{3+} in Trigonal Environments. *J. Chem. Phys.* **1965**, *42*, 1999–2010. [CrossRef]
37. Rai, R. Jahn-Teller Theory of the Ti^{3+} Ion in Corundum. *Phys. Status Solidi B* **1972**, *52*, 671–681. [CrossRef]
38. Sanchez, A.; Strauss, A.J.; Aggarwal, R.L.; Fahey, R.E. Crystal growth, spectroscopy, and laser characteristics of Ti:Al/sub 2/O/sub 3/. *IEEE J. Quantum Electron.* **1988**, *24*, 995–1002. [CrossRef]
39. Moulton, P.F.; Cederberg, J.G.; Stevens, K.T.; Foundos, G.; Koselja, M.; Preclikova, J. Characterization of absorption bands in Ti:sapphire crystals. *Opt. Mater. Express* **2019**, *9*, 2216–2251. [CrossRef]
40. Macfarlane, R.M.; Wong, J.Y.; Sturge, M.D. Dynamic Jahn-Teller Effect in Octahedrally Coordinated d^1 Impurity Systems. *Phys. Rev.* **1968**, *166*, 250–258. [CrossRef]
41. Bausá, L.E.; Vergara, I.; García-Solé, J.; Strek, W.; Deren, P.J. Laser-excited luminescence in Ti-doped $MgAl_2O_4$ spinel. *J. Appl. Phys.* **1990**, *68*, 736–740. [CrossRef]
42. Izumi, K.; Miyazaki, S.; Yoshida, S.; Mizokawa, T.; Hanamura, E. Optical properties of 3d transition-metal-doped $MgAl_2O_4$ spinels. *Phys. Rev. B* **2007**, *76*, 075111. [CrossRef]
43. Sato, T.; Shirai, M.; Tanaka, K.; Kawabe, Y.; Hanamura, E. Strong blue emission from Ti-doped $MgAl_2O_4$ crystals. *J. Lumin.* **2005**, *114*, 155–161. [CrossRef]
44. Jouini, A.; Yoshikawa, A.; Brenier, A.; Fukuda, T.; Boulon, G. Optical properties of transition metal ion-doped $MgAl_2O_4$ spinel for laser application. *Phys. Status Solidi C* **2007**, *4*, 1380–1383. [CrossRef]
45. Bausa, L.E.; Jaque, F.; Garcia Sole, J.; Duran, A. Photoluminescence of Ti^{3+} in P_2O_5-Na_2O-Al_2O_3 glass. *J. Mater. Sci.* **1988**, *23*, 1921–1922. [CrossRef]
46. Sun, G.H.; Zhang, Q.L.; Luo, J.Q.; Li, L.C.; Deng, Z.; Zhang, R.G. The effect of annealing on spectral characteristics of the Ti doped $MgAl_2O_4$ crystal. *J. Lumin.* **2021**, *234*, 117956. [CrossRef]
47. Grytsina, V.T.; Kazarinov, Y.G.; Kobyakov, V.A.; Lytvynov, L.A. Growth and Characterization of Titanium Doped Spinel Crystals. *Acta Phys. Pol. A* **2018**, *133*, 774–777. [CrossRef]
48. Meyn, J.-P.; Danger, T.; Petermann, K.; Huber, G. Spectroscopic characterization of V^{4+} -doped Al_2O_3 and $YAlO_3$. *J. Lumin.* **1993**, *55*, 55–62. [CrossRef]
49. Ballhausen, C.J. *Introduction to Ligand Field Theory*; McGraw-Hill: New York, NY, USA, 1962; ISBN 978-0-07-003580-5.
50. Kniec, K.; Marciniak, L. Spectroscopic Properties of Vanadium Ions for Applications in Luminescent Nanothermometry. In *Light-Matter Interactions Towards the Nanoscale*; Cesaria, M., Calà Lesina, A., Collins, J., Eds.; Springer: Dordrecht, The Netherlands, 2022; pp. 329–330.
51. Johnston, W.D. Optical Spectra of the Various Valence States of Vanadium in Na_2 $O.2SiO_2$ Glass. *J. Am. Ceram. Soc.* **1965**, *48*, 608–611. [CrossRef]
52. Liehr, A.D.; Ballhausen, C.J. Complete theory of Ni(II) and V(III) in cubic crystalline fields. *Ann. Phys.* **1959**, *6*, 134–155. [CrossRef]
53. Carman, C.J.; Kroenke, W.J. Electron spin resonance of.alpha.-chromia-alumina solid solutions. *J. Phys. Chem.* **1968**, *72*, 2562–2566. [CrossRef]
54. Caird, J.A. Chromium Activated Crystals as Tunable Laser Media—What Makes Them Special? In *Tunable Solid-State Lasers II*; Budgor, A.B., Esterowitz, L., DeShazer, L.G., Eds.; Springer: Berlin/Heidelberg, Germany, 1986; pp. 20–34.
55. Kusuma, H.H.; Astuti, B.; Ibrahim, Z. Absorption and emission properties of ruby (Cr:Al_2O_3) single crystal. *J. Phys. Conf. Ser.* **2019**, *1170*, 012054. [CrossRef]
56. Hunault, M.O.J.Y.; Harada, Y.; Miyawaki, J.; Wang, J.; Meijerink, A.; de Groot, F.M.F.; van Schooneveld, M.M. Direct Observation of Cr^{3+} 3d States in Ruby: Toward Experimental Mechanistic Evidence of Metal Chemistry. *J. Phys. Chem. A* **2018**, *122*, 4399–4413. [CrossRef]
57. Saraswati, V.; Ramarao, G.V. Optical properties of coloured transparent alumina gel monolith glasses. *Bull. Mater. Sci.* **1987**, *9*, 193–202. [CrossRef]
58. Maiman, T.H.; Hoskins, R.H.; D'Haenens, I.J.; Asawa, C.K.; Evtuhov, V. Stimulated Optical Emission in Fluorescent Solids. II. Spectroscopy and Stimulated Emission in Ruby. *Phys. Rev.* **1961**, *123*, 1151–1157. [CrossRef]
59. Zandonà, A.; Castaing, V.; Shames, A.I.; Helsch, G.; Deubener, J.; Becerro, A.I.; Allix, M.; Goldstein, A. Oxidation and coordination states assumed by transition metal dopants in an invert ultrabasic silicate glass. *J. Non-Cryst. Solids* **2023**, *603*, 122094. [CrossRef]
60. Pratapkumar, C.; Prashantha, S.C.; Nagabhushana, H.; Jnaneshwara, D.M. Photoluminescence and photometric studies of low temperature prepared red emitting $MgAl_2O_4$:Cr^{3+} nanophosphors for solid state displays. *J. Sci. Adv. Mater. Devices* **2018**, *3*, 464–470. [CrossRef]
61. Majewski-Napierkowski, A.; Gorbenko, V.; Zorenko, T.; Witkiewicz-Łukaszek, S.; Zorenko, Y. Regularities of Manganese Charge State Formation and Luminescent Properties of Mn-Doped Al_2O_3, YAlO3, and Y3Al5O12 Single Crystalline Films. *Crystals* **2023**, *13*, 1481. [CrossRef]

62. Ivakin, Y.D.; Danchevskaya, M.N.; Ovchinnikova, O.G.; Murav'eva, G.P.; Kreisberg, V.A. The kinetics and mechanism of doped corundum structure formation in an water fluid. *Russ. J. Phys. Chem. B* **2009**, *3*, 1019–1034. [CrossRef]
63. Kück, S. Laser-related spectroscopy of ion-doped crystals for tunable solid-state lasers. *Appl. Phys. B* **2001**, *72*, 515–562. [CrossRef]
64. Geschwind, S.; Kisliuk, P.; Klein, M.P.; Remeika, J.P.; Wood, D.L. Sharp-Line Fluorescence, Electron Paramagnetic Resonance, and Thermoluminescence of Mn^{4+} in α-Al_2O_3. *Phys. Rev.* **1962**, *126*, 1684–1686. [CrossRef]
65. Jovanić, B.R. Shift under pressure of the luminescence transitions of corundum doped with Mn^{4+}. *J. Lumin.* **1997**, *75*, 171–174. [CrossRef]
66. Katsumata, T.; Mitomi, H.; Nagayama, H.; Orihara, Y.; Aoki, M.; Yoshida, A.; Shiratake, K.; Minowa, S.; Sakuma, T.; Aizawa, H.; et al. Compositional variations in optical characteristics of Mn doped spinel crystals. *J. Cryst. Growth* **2017**, *468*, 387–390. [CrossRef]
67. Dutov, V.A.; Kiryakov, A.N.; Zatsepin, A.F.; Fokin, A.V.; Dyachkova, T.V.; Tytunnyuk, A.P.; Zainulin, Y.G. Impurity Mn^{2+} defects in $MgAl_2O_4$ nanoceramics. *AIP Conf. Proc.* **2020**, *2313*, 030013. [CrossRef]
68. Vink, A.P.; de Bruin, M.A.; Roke, S.; Peijzel, P.S.; Meijerink, A. Luminescence of Exchange Coupled Pairs of Transition Metal Ions. *J. Electrochem. Soc.* **2001**, *148*, E313. [CrossRef]
69. Bosi, F.; Hålenius, U.; Andreozzi, G.B.; Skogby, H.; Lucchesi, S. Structural refinement and crystal chemistry of Mn-doped spinel: A case for tetrahedrally coordinated Mn3+ in an oxygen-based structure. *Am. Mineral.* **2007**, *92*, 27–33. [CrossRef]
70. Ferguson, J.; Fielding, P.E. The origins of the colours of yellow, green and blue sapphires. *Chem. Phys. Lett.* **1971**, *10*, 262–265. [CrossRef]
71. Kai, A.; Matsuo, K. Electron spin resonance of transparent alumina ceramics fabricated by spark plasma sintering. *J. Am. Ceram. Soc.* **2018**, *101*, 4591–4597. [CrossRef]
72. Moon, A.R.; Phillips, M.R. Defect Clustering and Color in Fe,Ti: α-Al_2O_3. *J. Am. Ceram. Soc.* **1994**, *77*, 356–367. [CrossRef]
73. Lehmann, G. Interstitial incorporation of di- and trivalent cobalt in quartz. *J. Phys. Chem. Solids* **1969**, *30*, 395–399. [CrossRef]
74. Schuster, C.; Klimke, J.; Schwingenschlögl, U. High energy transmission of Al_2O_3 doped with light transition metals. *J. Chem. Phys.* **2012**, *136*, 044522. [CrossRef]
75. Eigenmann, K.; Günthard, H.H. Valence states, redox reactions and biparticle formation of Fe and Ti doped sapphire. *Chem. Phys. Lett.* **1972**, *13*, 58–61. [CrossRef]
76. Eigenmann, K.; Kurtz, K.; Günthard, H.H. The optical spectrum of α-Al_2O_3:Fe^{3+}. *Chem. Phys. Lett.* **1972**, *13*, 54–57. [CrossRef]
77. Nikolska, L.; Terekova, V.; Samoilovich, M. On the origin of natural sapphire color. *Phys. Chem. Min.* **1978**, *3*, 213–224. [CrossRef]
78. Fritsch, E.; Rossman, G. An update on the color in gems. Part 1: Introduction and colors caused by dispersed metal ions. *Gems Gemol.* **1987**, *23*, 126–139. [CrossRef]
79. White, E.A.D. Synthetic gemstones. *Q. Rev. Chem. Soc.* **1961**, *15*, 1–29. [CrossRef]
80. Bramley, R.; McCool, M.B. EPR of Fe^{3+} pairs in α-Al_2O_3. *J. Phys. C Solid State Phys.* **1976**, *9*, 1793. [CrossRef]
81. Burns, R.G. *Mineralogical Applications of Crystal Field Theory*, 2nd ed.; Cambridge University Press: Cambridge, UK, 1993; ISBN 978-0-521-43077-7.
82. Weyl, W.A. *Coloured Glasses*; Society of Glass Technology: Sheffield, UK, 1951.
83. Bamford, C.R. *Colour Generation and Control in Glass*; Elsevier Scientific Publishing Company: Amsterdam, The Netherlands, 1977; ISBN 978-0-444-41577-6.
84. Sherman, D.M. Molecular orbital (SCF-Xα-SW) theory of metal-metal charge transfer processes in minerals. *Phys. Chem. Miner.* **1987**, *14*, 355–363. [CrossRef]
85. Vercamer, V. Spectroscopic and Structural Properties of Fe in Silicate glasses. Ph.D. Thesis, University Marie et Pierre Curie Paris-VI, Paris, France, 2016.
86. De Souza, S.S.; Ayres, F.; Blak, A.R. Simulation models of defects in $MgAl_2O_4$:Fe^{2+}, Fe^{3+} spinels. *Radiat. Eff. Defects Solids* **2001**, *156*, 311–316. [CrossRef]
87. Sharma, T.; Arora, B. Effect of Fe doping on structural and electrical properties of nanocrystalline spinel synthetized by solution combustion method. *Int. J. Res. Adv. Technol.* **2015**, *3*, 58–62.
88. Dickson, B.L.; Smith, G. Low-temperature optical absorption and Moessbauer spectra of staurolite and spinel. *Can. Mineral.* **1976**, *14*, 206–215.
89. Slack, G.A. $FeAl_2O_4$ -$MgAl_2O_4$: Growth and Some Thermal, Optical, and Magnetic Properties of Mixed Single Crystals. *Phys. Rev.* **1964**, *134*, A1268–A1280. [CrossRef]
90. Gaffney, E.S. Spectra of Tetrahedral Fe^{2+} in $MgAl_2O_4$. *Phys. Rev. B* **1973**, *8*, 3484–3486. [CrossRef]
91. Taran, M.N.; Koch-Müller, M.; Langer, K. Electronic absorption spectroscopy of natural (Fe^{2+}, Fe^{3+})-bearing spinels of spinel s.s.-hercynite and gahnite-hercynite solid solutions at different temperatures and high-pressures. *Phys. Chem. Miner.* **2005**, *32*, 175–188. [CrossRef]
92. Mao, H.K.; Bell, P.M. Crystal-field effects in spinel: Oxidation states of iron and chromium. *Geochim. Cosmochim. Acta* **1975**, *39*, 865–874. [CrossRef]

93. Osipov, V.V.; Shitov, V.A.; Luk'yashin, K.E.; Platonov, V.V.; Solomonov, V.I.; Korsakov, A.S.; Medvedev, A.I. Synthesis and study of Fe^{2+}: $MgAl_2O_4$ ceramics for active elements of solid-state lasers. *Quantum Electron.* **2019**, *49*, 89. [CrossRef]
94. Basyrova, L.; Bukina, V.; Balabanov, S.; Belyaev, A.; Drobotenko, V.; Dymshits, O.; Alekseeva, I.; Tsenter, M.; Zapalova, S.; Khubetsov, A.; et al. Synthesis, structure and spectroscopy of Fe^{2+}:$MgAl_2O_4$ transparent ceramics and glass-ceramics. *J. Lumin.* **2021**, *236*, 118090. [CrossRef]
95. Dutt, B.V.; Hurrell, J.P.; Kröger, F.A. High-Temperature Defect Structure of Cobalt-Doped α-Alumina. *J. Am. Ceram. Soc.* **1975**, *58*, 420–427. [CrossRef]
96. Keig, G.A. Influence of the valence state of added impurity ions on the observed color in doped aluminum oxide single crystals. *J. Cryst. Growth* **1968**, *2*, 356–360. [CrossRef]
97. Gächter, B.F.; Blum, H.; Günthard, H.H. ESR spectra and magnetic constants of α-Al_2O_3:Co^{2+},H+ and α-Al_2O_3:Co^{2+},X+. *Chem. Phys.* **1973**, *1*, 45–54. [CrossRef]
98. Geusic, J. ESR of Co^{2+} doped alumina. *Bull. Am. Phys. Soc.* **1959**, *11*, 261–268.
99. Zverev, G.; Prokhorov, A. Electron Paramagnetic Resonance and Spin-Lattice Relaxation of Co^{2+} in Corundum. *Sov. Phys. JETP* **1961**, *12*, 41–45.
100. Zhu, Z.; Zhang, H.; Wang, Y.; Wu, W.; Wu, L.; Zeng, N.; Ren, H.; Xu, S.; Goodman, B.A.; Deng, W. Preparation and optical properties of high-quality green cobalt sapphires. *J. Lumin.* **2024**, *267*, 120354. [CrossRef]
101. Han, X.; Feng, X.; Kang, Y.; Guo, S.; Li, Y. Intervalence charge transfer transition of Co^{2+}-Ti^{4+} ion pair in α-Al_2O_3:Co,Ti crystal. *J. Alloys Compd.* **2018**, *768*, 1058–1063. [CrossRef]
102. Danchevskaya, M.N.; Ivakin, Y.D.; Muravieva, G.P.; Luchkov, I.V. Synthesis and doping of fine-crystalline corundum in sub- and supercritical conditions. *J. Phys. Conf. Ser.* **2008**, *121*, 082001. [CrossRef]
103. Townsend, M.G.; Hill, O.F. Tetravalent cobalt ion in α-Al_2O_3. *Trans. Faraday Soc.* **1965**, *61*, 2597–2602. [CrossRef]
104. Kokta, M.R.; Peressini, D.L.; Cooke, J.A.; Goodnight, K.L. Co-Doped Saturable Q-Switches and Laser Systems. U.S. Patent 2003/0007520 A1, 9 January 2003.
105. Chauviré, B.; Rondeau, B.; Fritsch, E.; Ressigeac, P.; Devidal, J.L. Blue Spinel from the Luc Yen District of Vietnam. *Gems Gemol.* **2015**, *51*, 2–17. [CrossRef]
106. Sardar, D.; Gruber, J.; Zandi, B.; Ferry, M.; Kokta, M. Spectroscopic properties of Co^{2+} in related spinels. *J. Appl. Phys.* **2002**, *91*, 4846–4852. [CrossRef]
107. Serment, B.; Brochon, C.; Hadziioannou, G.; Buffière, S.; Demourgues, A.; Gaudon, M. The versatile Co^{2+}/Co^{3+} oxidation states in cobalt alumina spinel: How to design strong blue nanometric pigments for color electrophoretic display. *RSC Adv.* **2019**, *9*, 34125–34135. [CrossRef] [PubMed]
108. D'Amario, L.; Föhlinger, J.; Boschloo, G.; Hammarström, L. Unveiling hole trapping and surface dynamics of NiO nanoparticles. *Chem. Sci.* **2018**, *9*, 223–230. [CrossRef] [PubMed]
109. Thomas, V.; Mashkovtsev, R.; Smirnov, S.; Maltsev, V. Tairus Hydrothermal Synthetic Sapphires Doped with Nickel and Chromium. *Gems Gemol.* **1997**, *33*, 188–202. [CrossRef]
110. Solntsev, V.P.; Tsvetkov, E.G.; Alimpiev, A.I.; Mashkovtsev, R.I. Coordination and valent state of nickel ions in beryl and chrysoberyl crystals. *Phys. Chem. Miner.* **2006**, *33*, 300–313. [CrossRef]
111. Brik, M.G. Crystal field analysis of the absorption spectra and electron–phonon interaction in Ca3Sc2Ge3O12:Ni^{2+}. *J. Phys. Chem. Solids* **2006**, *67*, 738–744. [CrossRef]
112. Sakurai, T.; Ishigame, M.; Arashi, H. Absorption Spectrum of Ni^{2+} Ions in Spinel. *J. Chem. Phys.* **1969**, *50*, 3241–3245. [CrossRef]
113. Blumberg, W.E.; Eisinger, J.; Geschwind, S. Cu^{3+} Ion in Corundum. *Phys. Rev.* **1963**, *130*, 900–909. [CrossRef]
114. Wei, Q.; Yang, Z.-Y.; Xu, Q.-M. Study on the local structure and the spin-Hamiltonian parameters for Cu^{3+} ions in Al_2O_3 crystal. *Phys. Status Solidi B* **2007**, *244*, 1997–2001. [CrossRef]
115. Petrosyan, A.K.; Mikzakiianyan, A.A. Zero-Field Splitting and g-Values of d^8 Ions in a Trigonal Crystal Field. *Phys. Status Solidi B* **1986**, *133*, 315–320. [CrossRef]
116. Feng, W.-L.; Zheng, W.-C.; Xue, J.-Y.; Zhao, F. Unified calculations of optical and EPR data for Cu^{3+} ion in Al_2O_3 crystal. *J. Magn. Magn. Mater.* **2015**, *377*, 190–192. [CrossRef]
117. Li, Y.-J.; Ye, S.; Wang, C.-H.; Wang, X.-M.; Zhang, Q.-Y. Temperature-dependent near-infrared emission of highly concentrated Cu^{2+} in $CaCuSi_4O_{10}$ phosphor. *J. Mater. Chem. C* **2014**, *2*, 10395–10402. [CrossRef]
118. Goldstein, A.; Vulfson, M.; Sirota, M. Optical Spectra of Copper-Doped Zn-Phosphate Glasses. *J. Am. Ceram. Soc.* **2007**, *90*, 3680–3682. [CrossRef]

119. Doremus, R.; Kao, S.-C.; Garcia, R. Optical absorption of small copper particles and the optical properties of copper. *Appl. Opt.* **1992**, *31*, 5773–5778. [CrossRef]
120. Jørgensen, C.K. The interelectronic repulsion and partly covalent bonding in transition-group complexes. *Discuss. Faraday Soc.* **1958**, *26*, 110–115. [CrossRef]

MDPI AG
Grosspeteranlage 5
4052 Basel
Switzerland
Tel.: +41 61 683 77 34

Ceramics Editorial Office
E-mail: ceramics@mdpi.com
www.mdpi.com/journal/ceramics